COMPUTATIONAL BIOLOGY

Edited by

Dr Ashwani Kumar

ArclerPress

Computational Biology

Edited by Dr Ashwani Kumar

Edition 2018

a

ArclerPress

Published by

Arcler Press LLC

708 3rd Avenue, 6th Floor

New York

NY 10017

United States of America

www.arclerpress.com

ISBN: 978-1-68094-536-2 (Hardback)

Printed in United States of America

Library of Congress Control Number: 2017933107

About the Editor

Dr Ashwani Kumar

Ashwani obtained his Master's degree in Clinical Pharmacology from Delhi University, India in 2007. Over the years, he has handled roles of increasing responsibilities including developing, reviewing, and approving medical and scientific literature and communications, e.g. manuscripts, and abstracts for the high indexed peer-reviewed journals; research proposal, posters, blogs, and articles. His major interest's areas include pharmacogenomics, onco-pharmacogenomics, and neuroscience. He is currently working as freelance healthcare consultant at Sydney, Australia.

Table of Contents

3 COMPUTATIONAL PHYLOGENETICS

4 COMPUTATIONAL NEUROSCIENCE

5 MOLECULAR CLONING

6 RESTRICTION MAPPING

7 MOLECULAR MODELLING

8 GENOMIC LIBRARIES

9 SEQUENCE ANALYSIS

List of Abbreviations

AIC	Akaike information criterion
BACs	Bacterial artificial chromosomes
BCI	brain–computer interface
BIC	Bayesian information criterion
CRUM	Computational Representational Understanding of Mind
CSAIL	Computer Science and Artificial Intelligence Laboratory
DNI	direct neural interface
EEG	electroencephalography
EMP	Earth Microbiome Project
ERP	Evoked Response Potential
GMO	genetically modified microorganisms
GOS	Global Ocean survey
HMP	Human Microbiome Project
ISMB	Intelligent Systems for Molecular Biology
kb	kilobase-pairs
LRT	likelihood ratio test
MMI	mind-machine interface
MP	Maximum parsimony
MRCA	most recent common ancestor
MSA	multiple sequence alignment
NCC	neural correlates of consciousness
NERF	Natural Extension Reference Frame
NSF	National Science Foundation

ORF	open reading frame
PACs	Partificial chromosomes ,
PCR	polymerase chain reaction
RTA	real-time analysis
SAV	Sequencing Analysis Viewer
SBML	Systems Biology Markup Language
TIGR	The Institute for Genomic Research
UCLA	University of California, Los Angeles
YACs	Yeast artificial chromosomes

Preface

Computational biology has soared from being an auxiliary discipline to being a crucial element for progress in practically all aspects of the biological sciences. Computational biology and bioinformatics is an interdisciplinary field that develops and applies computational methods to analyze large collections of biological data, such as genetic sequences, cell populations or protein samples, to make new predictions or discover new biology. The computational methods used include analytical methods, mathematical modelling and simulation. Prior to the advent of computational biology, biologists were unable to have access to large amounts of data. Researchers were able to develop analytical methods for interpreting biological information, but were unable to share them quickly among colleagues. Since the late 1990s, computational biology has become an important part of developing emerging technologies for the field of biology. The terms computational biology and evolutionary computation have a similar name, but are not to be confused. Unlike computational biology, evolutionary computation is not concerned with modeling and analyzing biological data. It instead creates algorithms based on the ideas of evolution across species. Sometimes referred to as genetic algorithms, the research of this field can be applied to computational biology. While evolutionary computation is not inherently a part of computational biology, Computational evolutionary biology is a subfield of it. Computational biology has been used to help sequence the human genome, create accurate models of the human brain, and assist in modeling biological systems.

Content Coverage

Chapters One and *Two* present an introduction of modelling biological systems and computational genomics, correspondingly. The analysis of biological systems refers to the understanding of biology and its mechanisms, along with the modelling of its dynamical behavior, whereas computational genomics refers to the use of computational and statistical analysis to decipher biology from genome sequences and related data, including both DNA and RNA sequence as well as other "post-genomic" data.

Chapters Three and *Four* present a coverage of computational phylogenetics and computational neuroscience, respectively. Computational phylogenetics aims to assemble a phylogenetic tree representing a hypothesis about the evolutionary ancestry of a set of genes, species, or other taxa, while computational neuroscience studies brain function in terms of the information processing properties of the structures that make up the nervous system.

Chapter Five is intended to focus on molecular cloning; a set of experimental methods in molecular biology that are used to assemble recombinant DNA molecules and to direct their replication within host organisms.

Chapter Six presents an overview of restriction mapping for obtaining structural information on a piece of DNA by the use of restriction enzymes. It is used to map an unknown segment of DNA by breaking it into pieces and identifying the locations of the breakpoints.

Chapters Seven and *Eight* focus on molecular modelling and genomic libraries, separately. Molecular modelling encompasses all methods, theoretical and computational, used to model or mimic the behavior of molecules, however genomic libraries are commonly used for sequencing applications.

Chapter Nine introduces sequence analysis, which is the process of subjecting a DNA, RNA or peptide sequence to any of a wide range of analytical methods to understand its features, function, structure, or evolution.

CHAPTER 1

Modelling Biological Systems

Chapter Contents

- **Introduction**
- **Overview**
- **Particular Tasks**
- **Ecosystem Model**

The analysis of biological systems refers to the understanding of biology and its mechanisms, along with the modelling of its dynamical behaviour. Mathematical observation and simulation is strictly linked to experimenting and aims to describe the functions and interactions of key elements of living systems.

INTRODUCTION

Modelling of biological systems is a significant task of systems biology and mathematical biology. A computational systems biology aims to develop and use efficient algorithms, data structures, visualization and communication tools with the goal of computer modelling of biological systems. It involves the use of computer simulations of biological systems, including cellular subsystems (such as the networks of metabolites and enzymes which comprise metabolism, signal transduction pathways and gene regulatory networks), to both analyze and visualize the complex connections of these cellular processes.

Artificial life or virtual evolution attempts to understand evolutionary processes via the computer simulation of simple (artificial) life forms.

OVERVIEW

It is understood that an unexpected emergent property of a complex system is a result of the interplay of the cause-and-effect among simpler, integrated parts (biological organisation). Biological systems manifest many important examples of emergent properties in the complex interplay of components. Traditional study of biological systems requires reductive methods in which quantities of data are gathered by category, such as concentration over time in response to a certain stimulus. Computers are critical to analysis and modelling of these data. The goal is to create accurate real-time models of a system's response to environmental and internal stimuli, such as a model of a cancer cell in order to find weaknesses in its signalling pathways, or modelling of ion channel mutations to see effects on cardiomyocytes and in turn, the function of a beating heart.

Standards

By far the most widely accepted standard format for storing and exchanging models in the field is the Systems Biology Markup Language (SBML) The SBML.org website includes a guide to many important software packages used in computational systems biology. Other markup languages with different emphases include BioPAX and CellML.

PARTICULAR TASKS

Cellular Model

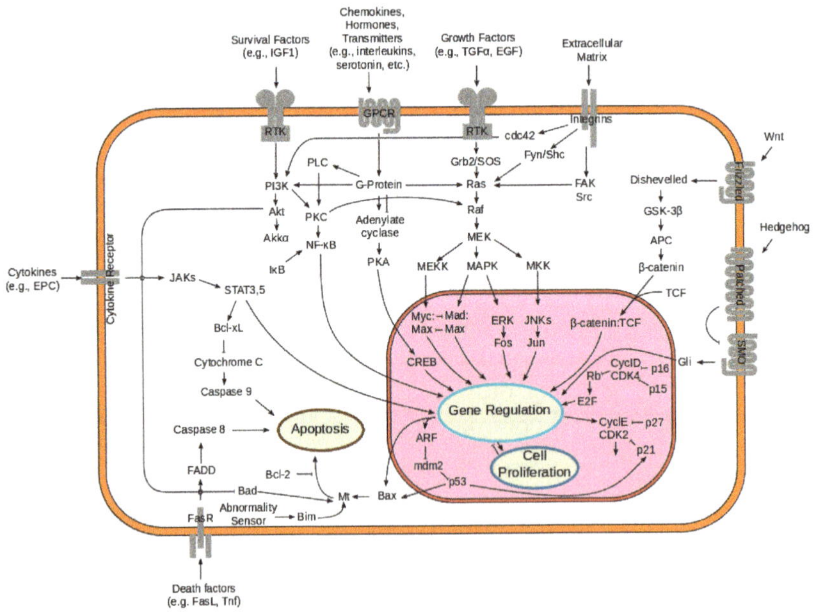

Part of the Cell Cycle

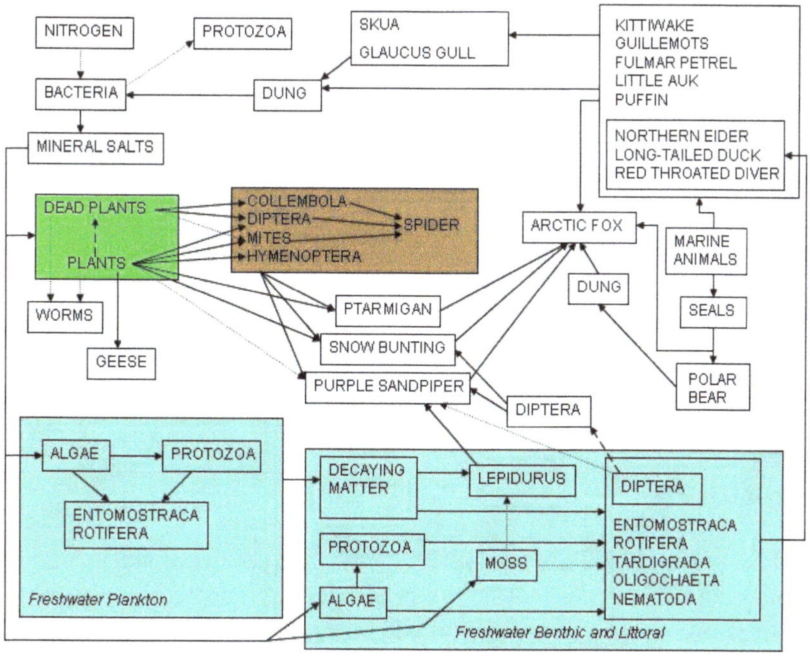

Summerhayes and Elton's 1923 food web of Bear Island (*Arrows represent an organism being consumed by another organism*).

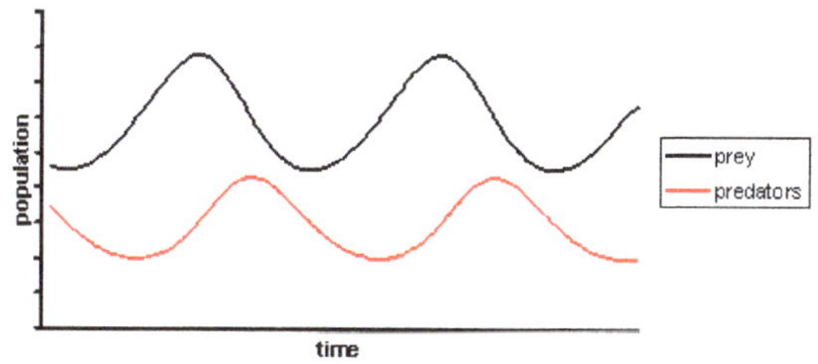

A sample time-series of the Lotka–Volterra model. Note that the two populations exhibit cyclic behaviour.

Creating a cellular model has been a particularly challenging task of systems biology and mathematical biology. It involves the use of computer simulations of the many cellular subsystems such as the networks of metabolites and enzymes which comprise metabolism, signal transduction pathways and gene regulatory networks to both analyze and visualize the complex connections of these cellular processes.

The complex network of biochemical reaction/transport processes and their spatial organization make the development of a predictive model of a living cell a grand challenge for the 21st century, listed as such by the National Science Foundation (NSF) in 2006.

A whole cell computational model for the bacterium *Mycoplasma genitalium*, including all its 525 genes, gene products, and their interactions, was built by scientists from Stanford University and the J. Craig Venter Institute and published on 20 July 2012 in Cell.

A dynamic computer model of intracellular signaling was the basis for Merrimack Pharmaceuticals to discover the target for their cancer medicine MM-111.

Membrane computing is the task of modelling specifically a cell membrane.

Multi-Cellular Organism Simulation

An open source simulation of C. elegans at the cellular level is being pursued by the OpenWorm community. So far the physics engine Gepetto has been built and models of the neural connectome and a muscle cell have been created in the NeuroML format.

Protein Folding

Protein structure prediction is the prediction of the three-dimensional structure of a protein from its amino acid sequence—that is, the prediction of a protein's tertiary structure from its primary structure. It is one of the most important goals pursued by bioinformatics and theoretical chemistry. Protein structure prediction is of high importance in medicine (for example, in drug design) and biotechnology (for example, in the design of novel enzymes). Every two years, the performance of current methods is assessed in the CASP experiment.

Human Biological Systems

Brain Model

The Blue Brain Project is an attempt to create a synthetic brain by reverse-engineering the mammalian brain down to the molecular level. The aim of this project, founded in May 2005 by the Brain and Mind Institute of the *École Polytechnique* in Lausanne, Switzerland, is to study the brain's architectural and functional principles. The project is headed by the Institute's director, Henry Markram. Using a Blue Gene supercomputer running Michael Hines›s NEURON software, the simulation does not consist simply of an artificial neural network, but involves a partially biologically realistic model of neurons. It is hoped by its proponents that it will eventually shed light on the nature of consciousness. There are a number of sub-projects, including the Cajal Blue Brain, coordinated by the Supercomputing and Visualization Center of Madrid (CeSViMa), and others run by universities and independent laboratories in the UK, U.S., and Israel. The Human Brain Project builds on the work of the Blue Brain Project. It is one of six pilot projects in the Future Emerging Technologies Research Program of the European Commission, competing for a billion euro funding.

Model of the Immune System

The last decade has seen the emergence of a growing number of simulations of the immune system.

Virtual Liver

The Virtual Liver project is a 43 million euro research program funded by the German Government, made up of seventy research group distributed across Germany. The goal is to produce a virtual liver, a dynamic mathematical model that represents human liver physiology, morphology and function.

Tree Model

Electronic trees (e-trees) usually use L-systems to simulate growth. L-systems are very important in the field of complexity science and A-life. A universally accepted system for describing changes in plant morphology at the cellular or modular level has yet to be devised. The most widely implemented tree generating algorithms are described in the papers "Creation and Rendering of Realistic Trees", and Real-Time Tree Rendering

ECOSYSTEM MODEL

An ecosystem model is an abstract, usually mathematical, representation of an ecological system (ranging in scale from an individual population, to an ecological community, or even an entire biome), which is studied to gain understanding of the real system.

Using data gathered from the field, ecological relationships — such as the relation of sunlight and water availability to photosynthetic rate, or that between predator and prey populations — are derived, and these are combined to form ecosystem models. These model systems are then studied in order to make predictions about the dynamics of the real system. Often, the study of inaccuracies in the model (when compared to empirical observations) will lead to the generation of hypotheses about possible ecological relations that are not yet known or well understood. Models enable researchers to simulate large-scale experiments that would be too costly or unethical to perform on a real ecosystem. They also enable the simulation of ecological processes over very long periods of time (i.e. simulating a process that takes centuries in reality, can be done in a matter of minutes in a computer model).

Ecosystem models have applications in a wide variety of disciplines, such as natural resource management, ecotoxicology and environmental health, agriculture, and wildlife conservation. It has also been applied to archaeology with varying degrees of success, through attempts to combine ecological models and archaeological models to explain mobility, diversity, etc. of stone tools

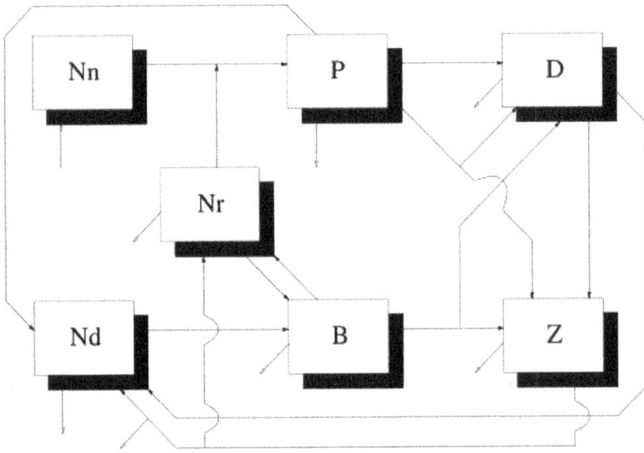

A structural diagram of the open ocean plankton ecosystem model of Fasham, Ducklow & McKelvie (1990).

Types of Models

There are two major types of ecological models, which are generally applied to different types of problems: (1) *analytic* models and (2) *simulation/computational* models. Analytic models are often more complex mathematically, and work best when dealing with relatively simple (often linear) systems, specifically those that can be accurately described by a set of mathematical equations whose behavior is well known. Simulation models on the other hand, use numerical techniques to solve problems for which analytic solutions are impractical or impossible. Simulation models tend to be more widely used, and are generally considered more ecologically realistic, while analytic models are valued for their mathematical elegance and explanatory power. Ecopath is a powerful software system which uses simulation and computational methods to model marine ecosystems. It is widely used by marine and fisheries scientists as a tool for modelling and visualising the complex relationships that exist in real world marine ecosystems.

Model Design

The process of model design begins with a specification of the problem to be solved, and the objectives for the model.

Ecological systems are composed of an enormous number of biotic and abiotic factors that interact with each other in ways that are often unpredictable, or so complex as to be impossible to incorporate into a computable model. Because of this complexity, ecosystem models typically simplify the systems they are studying to a limited number of components that are well understood, and deemed relevant to the problem that the model is intended to solve.

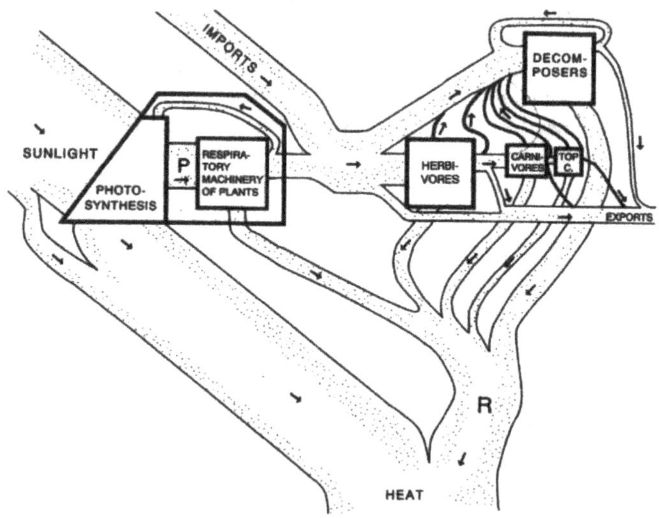

Diagram of the Silver Springs model (Odum, 1971). Note the aggregation into functional groups such as "herbivores" or "decomposers".

The process of simplification typically reduces an ecosystem to a small number of state variables and mathematical functions that describe the nature of the relationships between them. The number of ecosystem components that are incorporated into the model is limited by aggregating similar processes and entities into functional groups that are treated as a unit.

After establishing the components to be modeled and the relationships between them, another important factor in ecosystem model structure is the representation of space used. Historically, models have often ignored the confounding issue of space. However, for many ecological problems spatial dynamics are an important part of the problem, with different spatial environments leading to very different outcomes. *Spatially explicit models* (also called "spatially distributed" or "landscape" models) attempt to incorporate a heterogeneous spatial environment into the model. A spatial model is onc that has one or more state variables that are a function of space, or can be related to other spatial variables.

Validation

After construction, models are *validated* to ensure that the results are acceptably accurate or realistic. One method is to test the model with multiple sets of data that are independent of the actual system being studied. This is important since certain inputs can cause a faulty model to output correct results. Another method of validation is to compare the model's output with data collected from field observations. Researchers frequently specify beforehand how much of a disparity they are willing to accept between parameters output by a model and those computed from field data.

Example: The Lotka–Volterra Equations

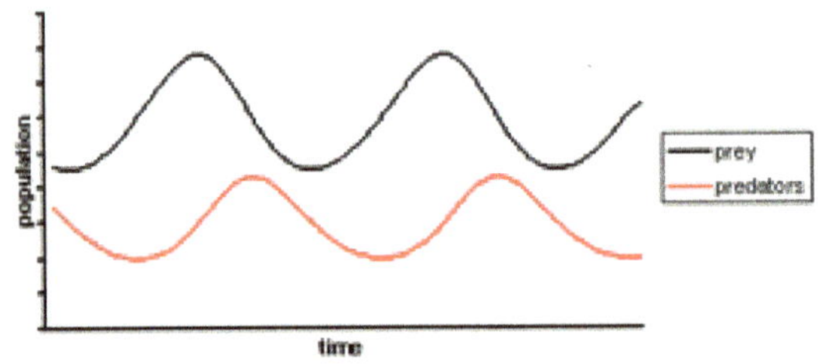

A sample time-series of the Lotka-Volterra model. Note that the two populations exhibit cyclic behaviour, and that the predator cycle lags behind that of the prey.

One of the earliest, and most well-known, ecological models is the predator-prey model of Alfred J. Lotka (1925) and Vito Volterra (1926). This model takes the form of a pair of ordinary differential equations, one representing a prey species, the other its predator.

$$\frac{dX}{dt} = \alpha . X - \beta . X . Y$$

$$\frac{dY}{dt} = \gamma . \beta . X . Y - \delta . Y$$

where,

- X is the number/concentration of the prey species;
- Y is the number/concentration of the predator species;
- α is the prey species' growth rate;
- β is the predation rate of Y upon X;
- γ is the assimilation efficiency of Y;
- δ is the mortality rate of the predator species

Volterra originally devised the model to explain fluctuations in fish and shark populations observed in the Adriatic Sea after the First World War (when fishing was curtailed). However, the equations have subsequently been applied more generally. Although simple, they illustrate some of the salient features of ecological models: modelled biological populations experience growth, interact with other populations (as either predators, prey or competitors) and suffer mortality.

A credible, simple alternative to the Lotka-Volterra predator-prey model and its common prey dependent generalizations is the ratio dependent or Arditi-Ginzburg model. The two are the extremes of the spectrum of predator interference models. According to the authors of the alternative view, the data show that true interactions in nature are so far from the Lotka-Volterra extreme on the interference spectrum that the model can simply be discounted as wrong. They are much closer to the ratio dependent extreme, so if a simple model is needed one can use the Arditi-Ginzburg model as the first approximation.

Models in Ecotoxicology

The purpose of models in ecotoxicology is the understanding, simulation and prediction of effects caused by toxicants in the environment. Most current models describe effects on one of many different levels of biological organization (e.g. organisms or populations). A challenge is the development of models that predict effects across biological scales. Ecotoxicology and models discusses some types of ecotoxicological models and provides links to many others.

Modelling of Infectious Disease

It is possible to model the progress of most infectious diseases mathematically to discover the likely outcome of an epidemic or to help manage them by vaccination. This field tries to find parameters for various infectious diseases and to use those parameters to make useful calculations about the effects of a mass vaccination programme.

REFERENCES

1. Andres Kriete, Roland Eils, Computational Systems Biology, Elsevier Academic Press, 2006.
2. Klipp, Liebermeister, Helbig, Kowald and Schaber. (2007). "Systems biology standards—the community speaks" (2007), Nature Biotechnology 25(4):390–391.
3. American Association for the Advancement of Science
4. Karr, J. (2012) A Whole-Cell Computational Model Predicts Phenotype from Genotype Cell
5. McDonagh, CF (2012) Antitumor Activity of a Novel Bispecific Antibody That Targets the ErbB2/ErbB3 Oncogenic Unit and Inhibits Heregulin-Induced Activation of ErbB3. Molecular Cancer Therapeutics
6. OpenWorm Downloads
7. Graham-Rowe, Duncan. "Mission to build a simulated brain begins", NewScientist, June 2005.
8. Palmer, Jason. Simulated brain closer to thought, BBC News.
9. The Human Brain Project. Archived July 5, 2012, at the Wayback Machine.
10. Video of Henry Markram presenting The Human Brain Project on 22 June 2012.
11. FET Flagships Initiative homepage.
12. Multi-criterion Evolutionary Algorithm with Model of the Immune System to Handle Constraints for Task Assignments - Springer
13. "Computer Simulation Captures Immune Response To Flu". Retrieved 2009-08-19.
14. Virtual Liver Network.
15. "Simulating plant growth". Archived from the original on 2009-12-09. Retrieved 2009-10-18.

16. Fasham, M. J. R.; Ducklow, H. W.; McKelvie, S. M. (1990). "A nitrogen-based model of plankton dynamics in the oceanic mixed layer". Journal of Marine Research. 48: 591–639. doi:10.1357/002224090784984678.
17. Hall, Charles A.S. & Day, John W. (1990). Ecosytem Modeling in Theory and Practice: An Introduction with Case Histories. University Press of Colorado. pp. 7–8. ISBN 0-87081-216-5.
18. Hall & Day, 1990: pp. 13-14
19. Dale, Virginia (2003). "Opportunities for Using Ecological Models for Resource Management". Ecological modeling for resource management. Springer. p. 3. ISBN 978-0-387-95493-6.
20. Pastorok, Robert A. (2002). "Introduction". Ecological modeling in risk assessment: chemical effects on populations, ecosystems, and landscapes. CRC Press. p. 7. ISBN 978-1-56670-574-5.
21. Forbes, Valery E. (2009). "The Role of Ecological Modeling in Risk Assessments Seen From an Academic's Point of View". In Thorbek, Pernille. Ecological Models for Regulatory Risk Assessments of Pesticides: Developing a Strategy for the Future. CRC Press. p. 89. ISBN 978-1-4398-0511-4.
22. Palladino, Paolo (1996). "Ecological Modeling and Integrated Pest Management". Entomology, ecology and agriculture: the making of scientific careers in North America, 1885-1985. Psychology Press. p. 153. ISBN 978-3-7186-5907-4.
23. Millspaugh, Joshua J. et al. (2008). "General Principles for Developing Landscape Models for Wildlife Conservation". Models for planning wildlife conservation in large landscapes. Academic Press. p. 1. ISBN 978-0-12-373631-4.
24. Marwick, Ben. Multiple Optima in Hoabinhian flaked stone artefact palaeoeconomics and palaeoecology at two archaeological sites in Northwest Thailand. Journal of Anthropological Archaeology 32.4 (2013): 553-564.
25. Jørgensen, Sven Erik (1996). Handbook of environmental and ecological modeling. CRC Press. pp. 403–404. ISBN 978-1-56670-202-7.
26. Grant, William Edward & Swannack, Todd M. (2008). Ecological modeling: a common-sense approach to theory and practice. John Wiley & Sons. p. 74. ISBN 978-1-4051-6168-8.
27. Hall, Charles A.S. & Day, John W. (1990). Ecosytem Modeling in Theory and Practice: An Introduction with Case Histories. University Press of Colorado. p. 9. ISBN 0-87081-216-5.
28. Pauly D, Christensen V, and Walters C (2000) Ecopath, Ecosim, and Ecospace as tools for evaluating ecosystem impact of fisheries ICES Journal of Marine Science, 57: 697–706. doi:10.1006/jmsc.2000.0726
29. Christensen V and Walters C (2004) Ecopath with Ecosim: methods, capabilities and limitations Ecological modelling, 172: 109–139.
30. Christensen V (2009) "The future of Ecopath" In: Palomares, MLD, Morissette L, Cisneros-Montemayor A, Varkey D, Coll M, Piroddi C (Eds), Ecopath 25

Years Conference Proceedings: Extended Abstracts, Fisheries Centre Research Reports 17(3): 159–160. University of British Columbia.

31. Odum, H.T. (1971). Environment, Power, and Society. Wiley-Interscience New York, N.Y.
32. Soetaert, Karline & Herman, Peter M.J. (2009). A practical guide to ecological modelling: using R as a simulation platform. Springer. p. 11. ISBN 978-1-4020-8623-6.
33. Gillman, Michael & Hails, Rosemary (1997). An introduction to ecological modelling: putting practice into theory. Wiley-Blackwell. p. 4. ISBN 978-0-632-03634-9.
34. Müller, Felix et al. (2011). "What are the General Conditions Under Which Ecological Models Can Be Applied". In Jopp, Fred et al. Modeling Complex Ecological Dynamics. Springer. pp. 13–14. ISBN 978-3-642-05028-2.
35. Hall & Day, 1990: p. 21
36. Hall & Day, 1990: p. 19
37. Buschke, F.T. & Seaman, M.T. 2011. Functional feeding groups as a taxonomic surrogate for a grassland arthropod assemblage. African Invertebrates 52 (1): 217-228.
38. McCallum, Hamish (2000). "Spatial Parameters". Population parameters: estimation for ecological models. Wiley-Blackwell. p. 184. ISBN 978-0-86542-740-2.
39. Tenhunen, John D. et al, eds. (2001). Ecosystem approaches to landscape management in Central Europe. Springer. pp. 586–587. ISBN 978-3-540-67267-8.
40. Ball, George L. (1999). "Ecological modeling". Encyclopedia of environmental science. Springer. p. 154. ISBN 978-0-412-74050-3.
41. Sklar, Fred H. & Hunsaker, Carolyn T. (2001). "The Use and Uncertainties of Spatial Data for Landscape Models: An Overview with Examples from the Florida Everglades". In Hunsaker, Carolyn T. Spatial uncertainty in ecology: implications for remote sensing and GIS applications. Springer. p. 15. ISBN 978-0-387-95129-4.
42. Jørgensen, Sven Erik & Bendoricchio, G. (2001). Fundamentals of ecological modelling. Gulf Professional Publishing. p. 79. ISBN 978-0-08-044028-6.
43. Pastorok, Robert A. (2002). "Introduction". Ecological modeling in risk assessment: chemical effects on populations, ecosystems, and landscapes. CRC Press. p. 22. ISBN 978-1-56670-574-5.
44. Shifley, S.R. (2008). "Validation of Landscape-Scale Decision Support Models That Predict Vegetation and Wildlife Dynamics". In Millspaugh, Joshua J. & Thompson, Frank Richard. Models for planning wildlife conservation in large landscapes. Academic Press. p. 419. ISBN 978-0-12-373631-4.
45. Voinov, Alexey (2008). Systems Science and Modeling for Ecological Economics. Academic Press. p. 131. ISBN 978-0-12-372583-7.

46. Reuter, Hauke et al. (2011). "How Valid Are Model Results? Assumptions, Validity Range and Documentation". In Jopp, Fred et al. Modeling Complex Ecological Dynamics. Springer. p. 325. ISBN 978-3-642-05028-2.
47. Earlier work on smallpox by Daniel Bernoulli and human overpopulation by Thomas Malthus predates that of Lotka and Volterra, but is not strictly ecological in nature
48. Lotka, A. J. (1925). The Elements of Physical Biology. Williams & Williams Co., Baltimore, USA.
49. Volterra, V. (1926). "Fluctuations in the abundance of a species considered mathematically". Nature. 118 (2983): 558–560. Bibcode:1926Natur.118..558V. doi:10.1038/118558a0.
50. Begon, M.; Harper, J. L.; Townsend, C. R. (1988). Ecology: Individuals, Populations and Communities. Blackwell Scientific Publications Inc., Oxford, UK.
51. Arditi, R. and Ginzburg, L.R. (1989) "Coupling in predator-prey dynamics: ratio dependence"Journal of Theoretical Biology, 139: 311–326.
52. Arditi, R. and Ginzburg, L.R. (2012) How Species Interact: Altering the Standard View on Trophic Ecology Oxford University Press. ISBN 9780199913831.

CHAPTER 2

Computational Genomics

Chapter Contents

Computational genomics includes: bio-sequence analysis, gene expression data analysis, phylogenetic analysis, and more specifically pattern recognition and analysis problems such as gene finding, motif finding, gene function prediction, fusion of sequence and expression information, and evolutionary models.

INTRODUCTION

Computational genomics (often referred to as Computational Genetics) refers to the use of computational and statistical analysis to decipher biology from genome sequences and related data, including both DNA and RNA sequence as well as other "post-genomic" data (i.e., experimental data obtained with technologies that require the genome sequence, such as genomic DNA microarrays). These, in combination with computational and statistical approaches to understanding the function of the genes and statistical association analysis, this field is also often referred to as Computational and Statistical Genetics/genomics. As such, computational genomics may be regarded as a subset of bioinformatics and computational biology, but with a focus on using whole genomes (rather than individual genes) to understand the principles of how the DNA of a species controls its biology at the molecular level and beyond. With the current abundance of massive biological datasets, computational studies have become one of the most important means to biological discovery.

The roots of computational genomics are shared with those of bioinformatics. During the 1960s, Margaret Dayhoff and others at the National Biomedical Research Foundation assembled databases of homologous protein sequences for evolutionary study. Their research developed a phylogenetic tree that determined the evolutionary changes that were required for a particular protein to change into another protein based on the underlying amino acid sequences. This led them to create a scoring matrix that assessed the likelihood of one protein being related to another. Beginning in the 1980s, databases of genome sequences began to be recorded, but this presented new challenges in the form of searching and comparing the databases of gene information. Unlike text-searching algorithms that are used on websites such as Google or Wikipedia, searching for sections of genetic similarity requires one to find strings that are not simply identical, but similar. This led to the development of the Needleman-Wunsch algorithm, which is a dynamic programming algorithm for comparing sets of amino acid sequences with each other by using scoring matrices derived from the earlier research by Dayhoff. Later, the BLAST algorithm was developed for performing fast, optimized searches of gene sequence databases. BLAST and its derivatives are probably the most widely used algorithms for this purpose.

The emergence of the phrase "computational genomics" coincides with the availability of complete sequenced genomes in the mid-to-late 1990s. The first meeting of the Annual Conference on Computational Genomics was organized by scientists from The Institute for Genomic Research (TIGR) in 1998, providing a forum for this speciality and effectively distinguishing this area of science from the more general fields of Genomics or Computational Biology. The first use of this term in scientific literature, according to MEDLINE abstracts, was just one year earlier in Nucleic Acids Research. The final Computational Genomics conference was held in 2006, featuring a keynote talk by Nobel Laureate Barry Marshall, co-discoverer of the link between Helicobacter pylori and stomach ulcers. As of 2014, the leading conferences in the field include Intelligent Systems for Molecular Biology (ISMB) and RECOMB.

The development of computer-assisted mathematics (using products such as Mathematica or Matlab) has helped engineers, mathematicians and computer scientists to start operating in this domain, and a public collection of case studies and demonstrations is growing, ranging from whole genome comparisons to gene expression analysis. This has increased the introduction of different ideas, including concepts from systems and control, information theory, strings analysis and data mining. It is anticipated that computational approaches will become and remain a standard topic for research and teaching, while students fluent in both topics start being formed in the multiple courses created in the past few years.

CONTRIBUTIONS OF COMPUTATIONAL GENOMICS RESEARCH TO BIOLOGY

Contributions of computational genomics research to biology include:

- discovering subtle patterns in genomic sequences
- proposing cellular signalling networks

- proposing mechanisms of genome evolution
- predict precise locations of all human genes using comparative genomics techniques with several mammalian and vertebrate species
- predict conserved genomic regions that are related to early embryonic development
- discover potential links between repeated sequence motifs and tissue-specific gene expression
- measure regions of genomes that have undergone unusually rapid evolution

Latest Development (From 2012)

First Computer Model of an Organism

Researchers at Stanford University created the first software simulation of an entire organism. The smallest free-living organism, *Mycoplasma genitalium*, has 525 genes which are fully mapped. With data from more than 900 scientific papers reported on the bacterium, researchers developed the software model using the object-oriented programming approach. A series of modules mimic the various functions of the cell and then are integrated together into a whole simulated organism. The simulation runs on a single CPU, recreates the complete life span of the cell at the molecular level, reproducing the interactions of molecules in cell processes including metabolism and cell division.

The 'silicon cell' will act as computerized laboratories that could perform experiments which are difficult to do on an actual organism, or could carry out procedures much faster. The applications will include faster screening of new compounds, understanding of basic cellular principles and behavior.

COMPUTATIONAL EPIGENETICS

Computational epigenetics unreliable source? uses bioinformatic methods clarification needed to complement experimental research in epigenetics. Due to the recent explosion of epigenome datasets, computational methods play an increasing role in all areas of epigenetic research.

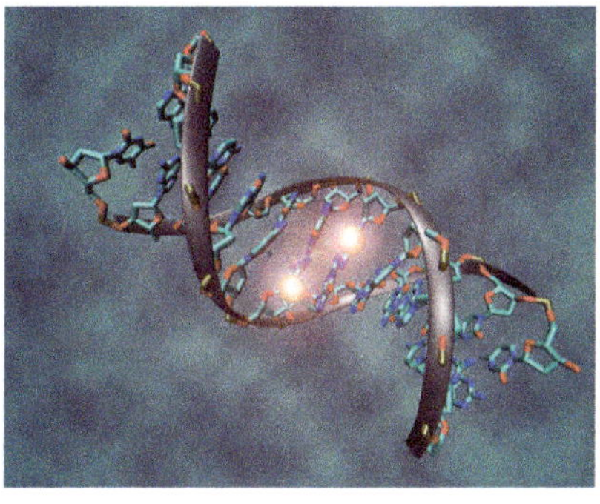

DNA methylation is a epigenetic mechanism that can be studied with bioinformatics

Definition

Research in computational epigenetics comprises the development and application of bioinformatics methods for solving epigenetic questions, as well as computational data analysis and theoretical modeling in the context of epigenetics. This includes modelling of the effects of histone and DNA CpG island methylation.

Current Research Areas

Epigenetic Data Processing and Analysis

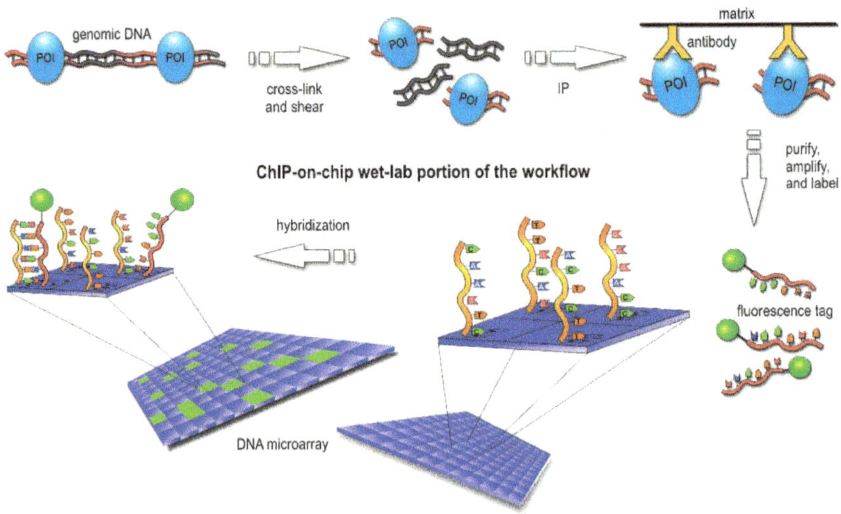

ChIP-on-chip technique

Various experimental techniques have been developed for genome-wide mapping of epigenetic information, the most widely used being ChIP-on-chip, ChIP-seq and bisulfite sequencing. All of these methods generate large amounts of data and require efficient ways of data processing and quality control by bioinformatic methods.

Epigenome Prediction

A substantial amount of bioinformatic research has been devoted to the prediction of epigenetic information from characteristics of the genome sequence. Such predictions serve a dual purpose. First, accurate epigenome predictions can substitute for experimental data, to some degree, which is particularly relevant for newly discovered epigenetic mechanisms and for species other than human and mouse. Second, prediction algorithms build statistical models of epigenetic information from training data and can therefore act as a first step toward quantitative modeling of an epigenetic mechanism. Successful computational prediction of DNA and lysine methylation and acetylation has been achieved by combinations of various features.

Applications in Cancer Epigenetics

The important role of epigenetic defects for cancer opens up new opportunities for improved diagnosis and therapy. These active areas of research give rise to two questions that are particularly amenable to bioinformatic analysis. First, given a list of genomic regions exhibiting epigenetic differences between tumor cells and controls (or between different disease subtypes), can we detect common patterns or find evidence of a functional relationship of these regions to cancer? Second, can we use bioinformatic methods in order to improve diagnosis and therapy by detecting and classifying important disease subtypes?

Emerging Topics

The first wave of research in the field of computational epigenetics was driven by rapid progress of experimental methods for data generation, which required adequate computational methods for data processing and quality control, prompted epigenome prediction studies as a means of understanding the genomic distribution of epigenetic information, and provided the foundation for initial projects on cancer epigenetics. While these topics will continue to be major areas of research and the mere quantity of epigenetic data arising from epigenome projects poses a significant bioinformatic challenge, several additional topics are currently emerging.

- Epigenetic regulatory circuitry: Reverse engineering the regulatory networks that read, write and execute epigenetic codes.
- Population epigenetics: Distilling regulatory mechanisms from the integration of epigenome data with gene expression profiles and haplotype maps for a large sample from a heterogeneous population.
- Evolutionary epigenetics: Learning about epigenome regulation in human (and its medical consequences) by cross-species comparisons.
- Theoretical modeling: Testing our mechanistic and quantitative understanding of epigenetic mechanisms by in silico simulation.
- Genome browsers: Developing a new blend of web services that enable biologists to perform sophisticated genome and epigenome analysis within an easy-to-use genome browser environment.
- Medical epigenetics: Searching for epigenetic mechanisms that play a role in diseases other than cancer, as there is strong circumstantial evidence for epigenetic regulation being involved in mental disorders, autoimmune diseases and other complex diseases.

Epigenetics Databases

- MethDB Contains information on 19,905 DNA methylation content data and 5,382 methylation patterns for 48 species, 1,511 individuals, 198 tissues and cell lines and 79 phenotypes.
- PubMeth Contains over 5,000 records on methylated genes in various cancer types.

- REBASE Contains over 22,000 DNA methyltransferases genes derived from GenBank.
- MeInfoText Contains gene methylation information across 205 human cancer types.
- MethPrimerDB Contains 259 primer sets from human, mouse and rat for DNA methylation analysis.
- The Histone Database Contains 254 sequences from histone H1, 383 from histone H2, 311 from histone H2B, 1043 from histone H3 and 198 from histone H4, altogether representing at least 857 species.
- ChromDB Contains 9,341 chromatin-associated proteins, including RNAi-associated proteins, for a broad range of organisms.
- CREMOFAC Contains 1725 redundant and 720 non-redundant chromatin-remodeling factor sequences in eukaryotes.
- The Krembil Family Epigenetics Laboratory Contains DNA methylation data of human chromosomes 21, 22, male germ cells and DNA methylation profiles in monozygotic and dizygotic twins.
- MethyLogiX DNA methylation database Contains DNA methylation data of human chromosomes 21 and 22, male germ cells and late-onset Alzheimer›s disease.

Sources and Further Reading

- The original version of this article was based on a review paper on computational epigenetics that appeared in the January 2008 issue of the Bioinformatics journal: *"Computational epigenetics". Bioinformatics. **24**: 1–10. January 2008. doi:10.1093/bioinformatics/btm546. PMID 18024971.*. This review paper provides >100 references to scientific papers and extensive background information.
- Additional data has been updated and added, based on a review paper on computational epigenetics that appeared in the January 2010 issue of the Bioinformation journal: *Lim S.J., Tan T.W. and Tong, J.C. (2010) Computational epigenetics: the new scientific paradigm. Bioinformation, 4(7): 331-337*. This review paper provides >129 references to scientific papers. It is published as open access and can be downloaded freely from the publisher's web page: http://bioinformation.net/004/007000042010.pdf.

REFERENCES

1. Koonin EV (March 2001). "Computational genomics". Curr. Biol. 11 (5): R155–8. doi:10.1016/S0960-9822(01)00081-1. PMID 11267880.
2. Computational Genomics and Proteomics at MIT
3. Mount, David (2000). Bioinformatics, Sequence and Genome Analysis. Cold Spring Harbor Laboratory Press. pp. 2–3. ISBN 0-87969-597-8.
4. Brown, T.A. (1999). Genomes. Wiley. ISBN 0-471-31618-0.

5. backPid=67&cHash=fd69079f5e The 7th Annual Conference on Computational Genomics (2004)
6. The 9th Annual Conference on Computational Genomics (2006)
7. Wagner A (September 1997). "A computational genomics approach to the identification of gene networks". Nucleic Acids Res. 25 (18): 3594–604. doi:10.1093/nar/25.18.3594. PMC 146952 . PMID 9278479.
8. Cristianini, N.; Hahn, M. (2006). Introduction to Computational Genomics. Cambridge University Press. ISBN 0-521-67191-4.
9. Gagniuc, P; Ionescu-Tirgoviste, C (Sep 28, 2012). "Eukaryotic genomes may exhibit up to 10 generic classes of gene promoters.". BMC Genomics. 13: 512. doi:10.1186/1471-2164-13-512. PMC 3549790 . PMID 23020586.
10. McClure, Max (19 July 2012). "Stanford researchers produce first complete computer model of an organism". Stanford University News. Retrieved 3 August 2012.
11. Karr JR, Sanghvi JC, Macklin DN, et al. (July 2012). "A whole-cell computational model predicts phenotype from genotype". Cell. 150 (2): 389–401. doi:10.1016/j.cell.2012.05.044. PMC 3413483 . PMID 22817898.
12. John Markoff (20 July 2012). "In First, Software Emulates Lifespan of Entire Organism". The New York Times. Retrieved 3 August 2012.
13. Bock, C; Lengauer T (2008). "Computational epigenetics". Bioinformatics. 24 (1): 1–10. doi:10.1093/bioinformatics/btm546. PMID 18024971.
14. Lim, S J; Tan T W; Tong J C (2010). "Computational epigenetics: the new scientific paradigm" (PDF). Bioinformation. 4 (7): 331–337. doi:10.6026/97320630004331.
15. Madrigal, P; Krajewski, P (July 2015). "Uncovering correlated variability in epigenomic datasets using the Karhunen-Loeve transform.". BioData Mining. 8: 20. doi:10.1186/s13040-015-0051-7.
16. Shi, Shao-Yun; Jian-Ding Qiu; Xing-Yu Sun; Sheng-Bao Suo; Shu-Yun Huang; Ru-Ping Liang (2012). "PLMLA: prediction of lysine methylation and lysine acetylation by combining multiple features". Mol. BioSyst. 8 (1): 1520–1527. doi:10.1039/C2MB05502C. PMID 22402705.
17. Zheng, Hao; Shi-Wen Jiang; Hongwei Wu (2011). "Enhancement on the Predictive Power of the Prediction Model for Human Genomic DNA Methylation" (PDF). BIOCOMP'11: The 2011 International Conference on Bioinformatics and Computational Biology.
18. Roznovat, I.A.; Ruskin H. J. (2013). "A Computational Model for Genetic and Epigenetic Signals in Colon Cancer". Interdisciplinary Sciences: Computational Life Sciences. 5 (3): 175–186. doi:10.1007/s12539-013-0172-y.
19. DNA Methylation Database
20. Pubmeth.Org
21. http://rebase.neb.com/rebase/rebase.html
22. MeInfoText: associated gene methylation and cancer information from text mining

23. methPrimerDB: the DNA methylation analysis PCR primer database
24. Histone Database - Histone Database
25. ChromDB::Chromatin Database
26. Cremofac
27. http://www.epigenomics.ca
28. Methylation Database

CHAPTER 3

Computational Phylogenetics

Chapter Contents

INTRODUCTION

Computational phylogenetics is the application of computational algorithms, methods, and programs to phylogenetic analyses. The goal is to assemble a phylogenetic tree representing a hypothesis about the evolutionary ancestry of a set of genes, species, or other taxa. For example, these techniques have been used to explore the family tree of hominid species and the relationships between specific genes shared by many types of organisms. Traditional phylogenetics relies on morphological data obtained by measuring and quantifying the phenotypic properties of representative organisms, while the more recent field of molecular phylogenetics uses nucleotide sequences encoding genes or amino acid sequences encoding proteins as the basis for classification. Many forms of molecular phylogenetics are closely related to and make extensive use of sequence alignment in constructing and refining phylogenetic trees, which are used to classify the evolutionary relationships between homologous genes represented in the genomes of divergent species. The phylogenetic trees constructed by computational methods are unlikely to perfectly reproduce the evolutionary tree that represents the historical relationships between the species being analyzed. The historical species tree may also differ from the historical tree of an individual homologous gene shared by those species.

Producing a phylogenetic tree requires a measure of homology among the characteristics shared by the taxa being compared. In morphological studies, this requires explicit decisions about which physical characteristics to measure and how to use them to encode distinct states corresponding to the input taxa. In molecular studies, a primary problem is in producing a multiple sequence alignment (MSA) between the genes or amino acid sequences of interest. Progressive sequence alignment methods produce a phylogenetic tree by necessity because they incorporate new sequences into the calculated alignment in order of genetic distance.

TYPES OF PHYLOGENETIC TREES AND NETWORKS

Phylogenetic trees generated by computational phylogenetics can be either *rooted* or *unrooted* depending on the input data and the algorithm used. A rooted tree is a directed graph that explicitly identifies a most recent common ancestor (MRCA), usually an imputed sequence that is not represented in the input. Genetic distance measures can be used to plot a tree with the input sequences as leaf nodes and their distances from the root proportional to their genetic distance from the hypothesized MRCA. Identification of a root usually requires the inclusion in the input data of at least one outgroup known to be only distantly related to the sequences of interest.

By contrast, unrooted trees plot the distances and relationships between input sequences without making assumptions regarding their descent. An unrooted tree can always be produced from a rooted tree, but a root cannot usually be placed on an unrooted tree without additional data on divergence rates, such as the assumption of the molecular clock hypothesis.

The set of all possible phylogenetic trees for a given group of input sequences can be conceptualized as a discretely defined multidimensional tree space through which search paths can be traced by optimization algorithms. Although counting the total number of trees for a nontrivial number of input sequences can be complicated by variations in the definition of a tree topology, it is always true that there are more rooted than unrooted trees for a given number of inputs and choice of parameters.

Both rooted and unrooted phylogenetic trees can be further generalized to rooted or unrooted phylogenetic networks, which allow for the modeling of evolutionary phenomena such as hybridization or horizontal gene transfer.

CODING CHARACTERS AND DEFINING HOMOLOGY

Morphological Analysis

The basic problem in morphological phylogenetics is the assembly of a matrix representing a mapping from each of the taxa being compared to representative measurements for each of the phenotypic characteristics being used as a classifier. The types of phenotypic data used to construct this matrix depend on the taxa being compared; for individual species, they may involve measurements of average body size, lengths or sizes of particular bones or other physical features, or even behavioral manifestations. Of course, since not every possible phenotypic characteristic could be measured and encoded for analysis, the selection of which features to measure is a major inherent obstacle to the method. The decision of which traits to use as a basis for the matrix necessarily represents a hypothesis about which traits of a species or higher taxon are evolutionarily relevant. Morphological studies can be confounded by examples of convergent evolution of phenotypes. A major challenge in constructing useful classes is the high likelihood of inter-taxon overlap in the distribution of the phenotype's variation. The inclusion of extinct taxa in morphological analysis is often difficult due to absence of or incomplete fossil records, but has been shown to have a significant effect on the trees produced; in one study only the inclusion of extinct species of apes produced a morphologically derived tree that was consistent with that produced from molecular data.

Some phenotypic classifications, particularly those used when analyzing very diverse groups of taxa, are discrete and unambiguous; classifying organisms as possessing or lacking a tail, for example, is straightforward in the majority of cases, as is counting features such as eyes or vertebrae. However, the most appropriate representation of continuously varying phenotypic measurements is a controversial problem without a general solution. A common method is simply to sort the measurements of interest into two or more classes, rendering continuous observed variation as discretely classifiable (e.g., all examples with humerus bones longer than a given cutoff are scored as members of one state, and all members whose humerus bones are shorter than the cutoff are scored as members of a second state). This results in an easily manipulated data set but has been criticized for poor reporting of the basis for the class definitions and for sacrificing information compared to methods that use a continuous weighted distribution of measurements. Because morphological data is extremely labor-intensive to collect, whether from literature sources or from field observations, reuse of previously compiled data matrices is not uncommon, although this may propagate flaws in the original matrix into multiple derivative analyses.

Molecular Analysis

The problem of character coding is very different in molecular analyses, as the characters in biological sequence data are immediate and discretely defined - distinct nucleotides in DNA or RNA sequences and distinct amino acids in protein sequences. However, defining homology can be challenging due to the inherent difficulties of multiple sequence alignment. For a given gapped MSA, several rooted phylogenetic trees can be constructed that vary in their interpretations of which changes are "mutations" versus ancestral characters, and which events are insertion mutations or deletion mutations. For example, given only a pairwise alignment with a gap region, it is impossible to determine whether one sequence bears an insertion mutation or the other carries a deletion. The problem is magnified in MSAs with unaligned and nonoverlapping gaps. In practice, sizable regions of a calculated alignment may be discounted in phylogenetic tree construction to avoid integrating noisy data into the tree calculation.

DISTANCE-MATRIX METHODS

Distance-matrix methods of phylogenetic analysis explicitly rely on a measure of "genetic distance" between the sequences being classified, and therefore they require an MSA as an input. Distance is often defined as the fraction of mismatches at aligned positions, with gaps either ignored or counted as mismatches. Distance methods attempt to construct an all-to-all matrix from the sequence query set describing the distance between each sequence pair. From this is constructed a phylogenetic tree that places closely related sequences under the same interior node and whose branch lengths closely reproduce the observed distances between sequences. Distance-matrix methods may produce either rooted or unrooted trees, depending on the algorithm used to calculate them. They are frequently used as the basis for progressive and iterative types of multiple sequence alignments. The main disadvantage of distance-matrix methods is their inability to efficiently use information about local high-variation regions that appear across multiple subtrees.

Neighbor-Joining

Neighbor-joining methods apply general cluster analysis techniques to sequence analysis using genetic distance as a clustering metric. The simple neighbor-joining method produces unrooted trees, but it does not assume a constant rate of evolution (i.e., a molecular clock) across lineages. Its relative, UPGMA (Unweighted Pair Group Method with Arithmetic mean) produces rooted trees and requires a constant-rate assumption - that is, it assumes an ultrametric tree in which the distances from the root to every branch tip are equal.

Fitch-Margoliash Method

The Fitch-Margoliash method uses a weighted least squares method for clustering based on genetic distance. Closely related sequences are given more weight in the tree construction process to correct for the increased inaccuracy in measuring distances between distantly related sequences. The distances used as input to the algorithm must be normalized to prevent large artifacts in computing relationships between closely related and distantly related groups. The distances calculated by this method must be linear; the linearity criterion for distances requires that the expected values of the branch lengths for two individual branches must equal the expected value of the sum of the two branch distances - a property that applies to biological sequences only when they have been corrected for the possibility of back mutations at individual sites. This correction is done through the use of a substitution matrix such as that derived from the Jukes-Cantor model of DNA evolution. The distance correction is only necessary in practice when the evolution rates differ among branches. Another modification of the algorithm can be helpful, especially in case of concentrated distances (please report to concentration of measure phenomenon and curse of dimensionality): that modification, described in, has been shown to improve the efficiency of the algorithm and its robustness.

The least-squares criterion applied to these distances is more accurate but less efficient than the neighbor-joining methods. An additional improvement that corrects for correlations between distances that arise from many closely related sequences in the data set can also be applied at increased computational cost. Finding the optimal least-squares tree with any correction factor is NP-complete, so heuristic search methods like those used in maximum-parsimony analysis are applied to the search through tree space.

Using Outgroups

Independent information about the relationship between sequences or groups can be used to help reduce the tree search space and root unrooted trees. Standard usage of distance-matrix methods involves the inclusion of at least one outgroup sequence known to be only distantly related to the sequences of interest in the query set. This usage can be seen as a type of experimental control. If the outgroup has been appropriately chosen, it will have a much greater genetic distance and thus a longer branch length than any other sequence, and it will appear near the root of a rooted tree. Choosing an appropriate outgroup requires the selection of a sequence that is moderately related to the sequences of interest; too close a relationship defeats the purpose of the outgroup and too distant adds noise to the analysis. Care should also be taken to avoid situations in which the species from which the sequences were taken are distantly related, but the gene encoded by the sequences is highly conserved across lineages. Horizontal gene transfer, especially between otherwise divergent bacteria, can also confound outgroup usage.

MAXIMUM PARSIMONY

Maximum parsimony (MP) is a method of identifying the potential phylogenetic tree that requires the smallest total number of evolutionary events to explain the observed sequence data. Some ways of scoring trees also include a "cost" associated with particular types of evolutionary events and attempt to locate the tree with the smallest total cost. This is a useful approach in cases where not every possible type of event is equally likely - for example, when particular nucleotides or amino acids are known to be more mutable than others.

The most naive way of identifying the most parsimonious tree is simple enumeration - considering each possible tree in succession and searching for the tree with the smallest score. However, this is only possible for a relatively small number of sequences or species because the problem of identifying the most parsimonious tree is known to be NP-hard; consequently a number of heuristic search methods for optimization have been developed to locate a highly parsimonious tree, if not the best in the set. Most such methods involve a steepest descent-style minimization mechanism operating on a tree rearrangement criterion.

Branch and Bound Algorithm

The branch and bound algorithm is a general method used to increase the efficiency of searches for near-optimal solutions of NP-hard problems first applied to phylogenetics in the early 1980s. Branch and bound is particularly well suited to phylogenetic tree construction because it inherently requires dividing a problem into a tree structure as it subdivides the problem space into smaller regions. As its name implies, it requires as input both a branching rule (in the case of phylogenetics, the addition of the next species or sequence to the tree) and a bound (a rule that excludes certain regions of the search space from consideration, thereby assuming that the optimal solution cannot occupy that region). Identifying a good bound is the most challenging aspect of the algorithm's application to phylogenetics. A simple way of defining the bound is a maximum number of assumed evolutionary changes allowed per tree. A set of criteria known as Zharkikh's rules severely limit the search space by defining characteristics shared by all candidate "most parsimonious" trees. The two most basic rules require the elimination of all but one redundant sequence (for cases where multiple observations have produced identical data) and the elimination of character sites at which two or more states do not occur in at least two species. Under ideal conditions these rules and their associated algorithm would completely define a tree.

Sankoff-Morel-Cedergren Algorithm

The Sankoff-Morel-Cedergren algorithm was among the first published methods to simultaneously produce an MSA and a phylogenetic tree for nucleotide sequences. The method uses a maximum parsimony calculation in conjunction with a scoring function that penalizes gaps and mismatches, thereby favoring the tree that introduces a minimal number of such events (an alternative view holds that the trees to be favored are those that maximize the amount of sequence similarity that can be interpreted as homology, a point of view that may lead to different optimal trees). The imputed sequences at the interior nodes of the tree are scored and summed over all the nodes in each possible tree. The lowest-scoring tree sum provides both an optimal tree and an optimal MSA given the scoring function. Because the method is highly computationally intensive, an approximate method in which initial guesses for the interior alignments are refined one node at a time. Both the full and the approximate version are in practice calculated by dynamic programming.

MALIGN and POY

More recent phylogenetic tree/MSA methods use heuristics to isolate high-scoring, but not necessarily optimal, trees. The MALIGN method uses a maximum-parsimony technique to compute a multiple alignment by maximizing a cladogram score, and its companion POY uses an iterative method that couples the optimization of the phylogenetic tree with improvements in the corresponding MSA. However, the use of these methods in constructing evolutionary hypotheses has been criticized as biased due to the deliberate construction of trees reflecting minimal evolutionary events. This, in turn, has been countered by the view that such methods should be seen as heuristic approaches to find the trees that maximize the amount of sequence similarity that can be interpreted as homology.

Maximum Likelihood

The maximum likelihood method uses standard statistical techniques for inferring probability distributions to assign probabilities to particular possible phylogenetic trees. The method requires a substitution model to assess the probability of particular mutations; roughly, a tree that requires more mutations at interior nodes to explain the observed phylogeny will be assessed as having a lower probability. This is broadly similar to the maximum-parsimony method, but maximum likelihood allows additional statistical flexibility by permitting varying rates of evolution across both lineages and sites. In fact, the method requires that evolution at different sites and along different lineages must be statistically independent. Maximum likelihood is thus well suited to the analysis of distantly related sequences, but it is believed to be computationally intractable to compute due to its NP-hardness.

The "pruning" algorithm, a variant of dynamic programming, is often used to reduce the search space by efficiently calculating the likelihood of subtrees. The method calculates the likelihood for each site in a "linear" manner, starting at a node whose only descendants are leaves (that is, the tips of the tree) and working backwards toward the "bottom" node in nested sets. However, the trees produced by the method are only rooted if the substitution model is irreversible, which is not generally true of biological systems. The search for the maximum-likelihood tree also includes a branch length optimization component that is difficult to improve upon algorithmically; general global optimization tools such as the Newton-Raphson method are often used.

BAYESIAN INFERENCE

Bayesian inference can be used to produce phylogenetic trees in a manner closely related to the maximum likelihood methods. Bayesian methods assume a prior probability distribution of the possible trees, which may simply be the probability of any one tree among all the possible trees that could be generated from the data, or may be a more sophisticated estimate derived from the assumption that divergence events such as speciation occur as stochastic processes. The choice of prior distribution is a point of contention among users of Bayesian-inference phylogenetics methods.

Implementations of Bayesian methods generally use Markov chain Monte Carlo sampling algorithms, although the choice of move set varies; selections used in Bayesian phylogenetics include circularly permuting leaf nodes of a proposed tree at each step and swapping descendant subtrees of a random internal node between two related trees. The use of Bayesian methods in phylogenetics has been controversial, largely due to incomplete specification of the choice of move set, acceptance criterion, and prior distribution in published work. Bayesian methods are generally held to be superior to parsimony-based methods; they can be more prone to long-branch attraction than maximum likelihood techniques,although they are better able to accommodate missing data.

Whereas likelihood methods find the tree that maximizes the probability of the data, a Bayesian approach recovers a tree that represents the most likely clades, by drawing on the posterior distribution. However, estimates of the posterior probability of clades (measuring their 'support') can be quite wide of the mark, especially in clades that aren't overwhelmingly likely. As such, other methods have been put forwards to estimate posterior probability.

MODEL SELECTION

Molecular phylogenetics methods rely on a defined substitution model that encodes a hypothesis about the relative rates of mutation at various sites along the gene or amino acid sequences being studied. At their simplest, substitution models aim to correct for differences in the rates of transitions and transversions in nucleotide sequences. The use of substitution models is necessitated by the fact that the genetic distance between two sequences increases linearly only for a short time after the two sequences diverge from each other (alternatively, the distance is linear only shortly before coalescence). The longer the amount of time after divergence, the more likely it becomes that two mutations occur at the same nucleotide site. Simple genetic distance calculations will thus undercount the number of mutation events that have occurred in evolutionary history. The extent of this undercount increases with increasing time since divergence, which can lead to the phenomenon of long branch attraction, or the misassignment of two distantly related but convergently evolving sequences as closely related. The maximum parsimony method is particularly susceptible to this problem due to its explicit search for a tree representing a minimum number of distinct evolutionary events.

Types of Models

All substitution models assign a set of weights to each possible change of state represented in the sequence. The most common model types are implicitly reversible because they assign the same weight to, for example, a G>C nucleotide mutation as to a C>G mutation. The simplest possible model, the Jukes-Cantor model, assigns an equal probability to every possible change of state for a given nucleotide base. The rate of change between any two distinct nucleotides will be one-third of the overall substitution rate. More advanced models distinguish between transitions and transversions. The most general possible time-reversible model, called the GTR model, has six mutation rate parameters. An even more generalized model known as the general 12-parameter model breaks time-reversibility, at the cost of much additional complexity in calculating genetic distances that are consistent among multiple lineages. One possible variation on this theme adjusts the rates so that overall GC content - an important measure of DNA double helix stability - varies over time.

Models may also allow for the variation of rates with positions in the input sequence. The most obvious example of such variation follows from the arrangement of nucleotides in protein-coding genes into three-base codons. If the location of the open reading frame (ORF) is known, rates of mutation can be adjusted for position of a given site within a codon, since it is known that wobble base pairing can allow for higher mutation rates in the third nucleotide of a given codon without affecting the codon's meaning in the genetic code. A less hypothesis-driven example that does not rely on ORF identification simply assigns to each site a rate randomly drawn from a predetermined distribution, often the gamma distribution or log-normal distribution. Finally, a more conservative estimate of rate variations known as the covarion method allows autocorrelated variations in rates, so that the mutation rate of a given site is correlated across sites and lineages.

Choosing the Best Model

The selection of an appropriate model is critical for the production of good phylogenetic analyses, both because underparameterized or overly restrictive models may produce aberrant behavior when their underlying assumptions are violated, and because overly complex or overparameterized models are computationally expensive and the parameters may be overfit.The most common method of model selection is the likelihood ratio test (LRT), which produces a likelihood estimate that can be interpreted as a measure of "goodness of fit" between the model and the input data. However, care must be taken in using these results, since a more complex model with more parameters will always have a higher likelihood than a simplified version of the same model, which can lead to the naive selection of models that are overly complex. For this reason model selection computer programs will choose the simplest model that is not significantly worse than more complex substitution models. A significant disadvantage of the LRT is the necessity of making a series of pairwise comparisons between models; it has been shown that the order in which the models are compared has a major effect on the one that is eventually selected.

An alternative model selection method is the Akaike information criterion (AIC), formally an estimate of the Kullback–Leibler divergence between the true model and the model being tested. It can be interpreted as a likelihood estimate with a correction factor to penalize overparameterized models. The AIC is calculated on an individual model rather than a pair, so it is independent of the order in which models are assessed. A related alternative, the Bayesian information criterion (BIC), has a similar basic interpretation but penalizes complex models more heavily.

A comprehensive step-by-step protocol on constructing phylogenetic tree, including DNA/Amino Acid contiguous sequence assembly, multiple sequence alignment, model-test (testing best-fitting substitution models) and phylogeny reconstruction using Maximum Likelihood and Bayesian Inference, is available at Nature Protocol

An non traditional way of evaluating the Pylogenetic Tree is to compare it with clustering result. One can use a Multidimensional Scaling technique, so called Interpolative Joining to do dimensionality reduction to visualize the clustering result for the sequences in 3D, and then map the phylogenetic tree onto the clustering result. A better tree usually has a higher correlation with the clustering result.

EVALUATING TREE SUPPORT

As with all statistical analysis, the estimation of phylogenies from character data requires an evaluation of confidence. A number of methods exist to test the amount of support for a phylogenetic tree, either by evaluating the support for each sub-tree in the phylogeny (nodal support) or evaluating whether the phylogeny is significantly different from other possible trees (alternative tree hypothesis tests).

Nodal Support

The most common method for assessing tree support is to evaluate the statistical support for each node on the tree. Typically, a node with very low support is not considered valid in further analysis, and visually may be collapsed into a polytomy to indicate that relationships within a clade are unresolved.

Consensus Tree

Many methods for assessing nodal support involve consideration of multiple phylogenies. The consensus tree summarizes the nodes that are shared among a set of trees. In a *strict consensus,* only nodes found in every tree are shown, and the rest are collapsed into an unresolved polytomy. Less conservative methods, such as the *majority-rule consensus* tree, consider nodes that are supported by a given percentage of trees under consideration (such as at least 50%).

For example, in maximum parsimony analysis, there may be many trees with the same parsimony score. A strict consensus tree would show which nodes are found in all equally parsimonious trees, and which nodes differ. Consensus trees are also used to evaluate support on phylogenies reconstructed with Bayesian inference (see below).

Bootstrapping and Jackknifing

In statistics, the bootstrap is a method for inferring the variability of data that has an unknown distribution using pseudoreplications of the original data. For example, given a set of 100 data points, a pseudoreplicate is a data set of the same size (100 points) randomly sampled from the original data, with replacement. That is, each original data point may be represented more than once in the pseudoreplicate, or not at all. Statistical support involves evaluation of whether the original data has similar properties to a large set of pseudoreplicates.

In phylogenetics, bootstrapping is conducted using the columns of the character matrix. Each pseudoreplicate contains the same number of species (rows) and characters (columns) randomly sampled from the original matrix, with replacement. A phylogeny is reconstructed from each pseudoreplicate, with the same methods used to reconstruct the phylogeny from the original data. For each node on the phylogeny, the nodal support is the percentage of pseudoreplicates containing that node.

The statistical rigor of the bootstrap test has been empirically evaluated using viral populations with known evolutionary histories, finding that 70% bootstrap support corresponds to a 95% probability that the clade exists. However, this was tested under ideal conditions (e.g. no change in evolutionary rates, symmetric phylogenies). In practice, values above 70% are generally supported and left to the researcher or reader to evaluate confidence. Nodes with support lower than 70% are typically considered unresolved.

Jackknifing in phylogenetics is a similar procedure, except the columns of the matrix are sampled without replacement. Pseudoreplicates are generated by randomly subsampling the data—for example, a 10% jackknife would involve randomly sampling 10% of the matrix many times to evaluate nodal support.

Posterior Probability

Reconstruction of phylogenies using Bayesian inference generates a posterior distribution of highly probable trees given the data and evolutionary model, rather than a single "best" tree. The trees in the posterior distribution generally have many different topologies.

Most Bayesian inference methods utilize a Markov-chain Monte Carlo iteration, and the initial steps of this chain are not considered reliable reconstructions of the phylogeny. Trees generated early in the chain are usually discarded as burn-in. The most common method of evaluating nodal support in a Bayesian phylogenetic analysis is to calculate the percentage of trees in the posterior distribution (post-burn-in) which contain the node.

The statistical support for a node in Bayesian inference is expected to reflect the probability that a clade really exists given the data and evolutionary model. Therefore, the threshold for accepting a node as supported is generally higher than for bootstrapping.

Step Counting Methods

Bremer support counts the number of extra steps needed to contradict a clade.

Shortcomings

These measures each have their weaknesses. For example, smaller or larger clades tend to attract larger support values than mid-sized clades, simply as a result of the number of taxa in them.

Bootstrap support can provide high estimates of node support as a result of noise in the data rather than the true existence of a clade.

LIMITATIONS AND WORKAROUNDS

Ultimately, there is no way to measure whether a particular phylogenetic hypothesis is accurate or not, unless the true relationships among the taxa being examined are already known (which may happen with bacteria or viruses under laboratory conditions). The best result an empirical phylogeneticist can hope to attain is a tree with branches that are well supported by the available evidence. Several potential pitfalls have been identified:

Homoplasy

Certain characters are more likely to evolve convergently than others; logically, such characters should be given less weight in the reconstruction of a tree. Weights in the form of a model of evolution can be inferred from sets of molecular data, so that maximum likelihood or Bayesian methods can be used to analyze them. For molecular sequences, this problem is exacerbated when the taxa under study have diverged substantially. As time since the divergence of two taxa increase, so does the probability of multiple substitutions on the same site, or back mutations, all of which result in homoplasies. For morphological data, unfortunately, the only objective way to determine convergence is by the construction of a tree – a somewhat circular method. Even so, weighting homoplasious characters does indeed lead to better-supported trees. Further refinement can be brought by weighting changes in one direction higher than changes in another; for instance, the presence of thoracic wings almost guarantees placement among the pterygote insects because, although wings are often lost secondarily, there is no evidence that they have been gained more than once.

Horizontal Gene Transfer

In general, organisms can inherit genes in two ways: vertical gene transfer and horizontal gene transfer. Vertical gene transfer is the passage of genes from parent to offspring, and horizontal (also called lateral) gene transfer occurs when genes jump between unrelated organisms, a common phenomenon especially in prokaryotes; a good example of this is the acquired antibiotic resistance as a result of gene exchange between various bacteria leading to multi-drug-resistant bacterial species. There have also been well-documented cases of horizontal gene transfer between eukaryotes.

Horizontal gene transfer has complicated the determination of phylogenies of organisms, and inconsistencies in phylogeny have been reported among specific groups of organisms depending on the genes used to construct evolutionary trees. The only way to determine which genes have been acquired vertically and which horizontally is to parsimoniously assume that the largest set of genes that have been inherited together have been inherited vertically; this requires analyzing a large number of genes.

Hybrids, Speciation, Introgressions and Incomplete Lineage Sorting

The basic assumption underlying the mathematical model of cladistics is a situation where species split neatly in bifurcating fashion. While such an assumption may hold on a larger scale (bar horizontal gene transfer, see above), speciation is often much less orderly. Research since the cladistic method was introduced has shown that hybrid speciation, once thought rare, is in fact quite common, particularly in plants. Also paraphyletic speciation is common, making the assumption of a bifurcating pattern unsuitable, leading to phylogenetic networks rather than trees. Introgression can also move genes between otherwise distinct species and sometimes even genera, complicating phylogenetic analysis based on genes. This phenomenon can contribute to "incomplete lineage sorting" and is thought to be a common phenomenon across a number of groups. In species level analysis this can be dealt with by larger sampling or better whole genome analysis. Often the problem is avoided by restricting the analysis to fewer, not closely related specimen.

Taxon Sampling

Owing to the development of advanced sequencing techniques in molecular biology, it has become feasible to gather large amounts of data (DNA or amino acid sequences) to infer phylogenetic hypotheses. For example, it is not rare to find studies with character matrices based on whole mitochondrial genomes (~16,000 nucleotides, in many animals). However, simulations have shown that it is more important to increase the number of taxa in the matrix than to increase the number of characters, because the more taxa there are, the more accurate and more robust is the resulting phylogenetic tree. This may be partly due to the breaking up of long branches.

Phylogenetic Signal

Another important factor that affects the accuracy of tree reconstruction is whether the data analyzed actually contain a useful phylogenetic signal, a term that is used generally to denote whether a character evolves slowly enough to have the same state in closely related taxa as opposed to varying randomly. Tests for phylogenetic signal exist.

Continuous Characters

Morphological characters that sample a continuum may contain phylogenetic signal, but are hard to code as discrete characters. Several methods have been used, one of which is gap coding, and there are variations on gap coding. In the original form of gap coding: group means for a character are first ordered by size. The pooled within-group standard deviation is calculated ... and differences between adjacent means ... are compared relative to this standard deviation. Any pair of adjacent means is considered different and given different integer scores ... if the means are separated by a "gap" greater than the within-group standard deviation ... times some arbitrary constant.

If more taxa are added to the analysis, the gaps between taxa may become so small that all information is lost. Generalized gap coding works around that problem by comparing individual pairs of taxa rather than considering one set that contains all of the taxa.

Missing Data

In general, the more data that are available when constructing a tree, the more accurate and reliable the resulting tree will be. Missing data are no more detrimental than simply having fewer data, although the impact is greatest when most of the missing data are in a small number of taxa. Concentrating the missing data across a small number of characters produces a more robust tree.

The Role of Fossils

Because many characters involve embryological, or soft-tissue or molecular characters that (at best) hardly ever fossilize, and the interpretation of fossils is more ambiguous than that of living taxa, extinct taxa almost invariably have higher proportions of missing data than living ones. However, despite these limitations, the inclusion of fossils is invaluable, as they can provide information in sparse areas of trees, breaking up long branches and constraining intermediate character states; thus, fossil taxa contribute as much to tree resolution as modern taxa. Fossils can also constrain the age of lineages and thus demonstrate how consistent a tree is with the stratigraphic record; stratocladistics incorporates age information into data matrices for phylogenetic analyses.

REFERENCES

1. Strait, DS; Grine, FE (2004). "Inferring hominoid and early hominid phylogeny using craniodental characters: the role of fossil taxa". J Hum Evol. 47 (6): 399–452. doi:10.1016/j.jhevol.2004.08.008.
2. Hodge, T; Cope, MJ (2000). "A myosin family tree". J Cell Sci. 113: 3353–3354.
3. Mount DM. (2004). Bioinformatics: Sequence and Genome Analysis 2nd ed. Cold Spring Harbor Laboratory Press: Cold Spring Harbor, New York.
4. Felsenstein J. (2004). Inferring Phylogenies Sinauer Associates: Sunderland, Massachusetts.
5. Swiderski DL, Zelditch ML, Fink WL. (1998). Why morphometrics is not special: coding quantitative data for phylogenetic analysis. 47(3):508-19.
6. Gaubert, P; Wozencraft, WC; Cordeiro-Estrela, P; Veron, G (2005). "Mosaics of convergences and noise in morphological phylogenies: what's in a viverrid-like carnivoran?". Syst Biol. 54 (6): 865–94. doi:10.1080/10635150500232769.
7. Wiens, JJ (2001). "Character analysis in morphological phylogenetics: problems and solutions". Syst Biol. 50 (5): 689–99. doi:10.1080/106351501753328811.
8. Jenner, RA (2001). "Bilaterian phylogeny and uncritical recycling of morphological data sets". Syst Biol. 50 (5): 730–743. doi:10.1080/106351501753328857.
9. Fitch, WM; Margoliash, E. (1967). "Construction of phylogenetic trees". Science. 155: 279–84. doi:10.1126/science.155.3760.279.
10. Lespinats, S.; Grando, D.; Maréchal, E.; Hakimi, M.A.; Tenaillon, O.; Bastien, O. (2011). "How Fitch-Margoliash Algorithm can benefit from Multi Dimensional Scaling". Evolutionary Bioinformatics. 7: 61–85.
11. Day, WHE (1987). "Computational complexity of inferring phylogenies from dissimilarity matrices". Bulletin of Mathematical Biology. 49 (4): 461–7. doi:10.1016/s0092-8240(87)80007-1.
12. Hendy, MD; Penny, D (1982). "Branch and bound algorithms to determine minimal evolutionary trees". Math Biosci. 60: 133–42.
13. Ratner VA, Zharkikh AA, Kolchanov N, Rodin S, Solovyov S, Antonov AS. (1995). Molecular Evolution Biomathematics Series Vol 24. Springer-Verlag: New York.
14. Sankoff, D; Morel, C; Cedergren, RJ (1973). "Evolution of 5S RNA and the non-randomness of base replacement". Nature New Biology. 245: 232–4. doi:10.1038/newbio245232a0.
15. De Laet J. (2005). Parsimony and the problem of inapplicables in sequence data. Pp. 81-116 in Albert, V.A. (ed.) Parsimony, phylogeny and genomics. Oxford University Press, ISBN 0-19-856493-7
16. Wheeler, WC; Gladstein, DG (1994). "MALIGN: a multiple nucleic acid sequence alignment program". J Heredity. 85: 417–18.
17. Simmons, MP (2004). "Independence of alignment and tree search". Mol Phylogenet Evol. 31 (3): 874–9.

18. De Laet, Jan (2015). "Parsimony analysis of unaligned sequence data: maximization of homology and minimization of homoplasy, not Minimization of operationally defined total cost or minimization of equally weighted transformations". Cladistics. doi:10.1111/cla.12098.

19. Chor B, Tuller T. (2005). Maximum likelihood of evolutionary trees: hardness and approximation. Bioinformatics 21.suppl 1: i97-i106.

20. Mau, B; Newton, MA (1997). "Phylogenetic inference for binary data on dendrograms using Markov chain Monte Carlo". J Comp Graph Stat. 6: 122–31. doi:10.2307/1390728.

21. Yang, Z; Rannala, B (1997). "bayesian phylogenetic inference using DNA sequences: a Markov chain Monte Carlo method". Mol Biol Evol. 46: 409–18.

22. Kolaczkowski, B.; Thornton, J. W. (2009). Delport, Wayne, ed. "Long-Branch Attraction Bias and Inconsistency in Bayesian Phylogenetics". PLoS ONE. 4 (12): e7891. doi:10.1371/journal.pone.0007891. PMC 2785476 . PMID 20011052.

23. Simmons, M. P. (2012). "Misleading results of likelihood-based phylogenetic analyses in the presence of missing data". Cladistics. 28 (2): 208–222. doi:10.1111/j.1096-0031.2011.00375.x.

24. Larget, B. (2013). "The Estimation of Tree Posterior Probabilities Using Conditional Clade Probability Distributions". Systematic Biology. 62 (4): 501–511. doi:10.1093/sysbio/syt014. PMC 3676676 . PMID 23479066.

25. Sullivan, Jack; Joyce, Paul (2005). "Model Selection in Phylogenetics". Annual Review of Ecology, Evolution, and Systematics. 36 (1): 445–466. doi:10.1146/annurev.ecolsys.36.102003.152633.

26. Galtier N, Guoy M. (1998.) Inferring pattern and process: maximum-likelihood implementation of a nonhomogeneous model of DNA sequence evolution for phylogenetic analysis. Mol. Biol. Evol. 15:871–79.

27. Fitch, WM; Markowitz, E (1970). "An improved method for determining codon variability in a gene and its application to the rate of fixation of mutations in evolution". Biochemical Genetics. 4: 579–593. doi:10.1007/bf00486096.

28. Pol D. (2004.) Empirical problems of the hierarchical likelihood ratio test for model selection. Syst Biol 53:949–62.

29. Bast, F. 2013. Sequence Similarity Search, Multiple Sequence Alignment, Model Selection, Distance Matrix and Phylogeny Reconstruction. Nature Protocol Exchange. doi: 10.1038/protex.2013.065

30. Ruan, Yang, et al. "Integration of Clustering and Multidimensional Scaling to Determine Phylogenetic Trees as Spherical Phylograms Visualized in 3 Dimensions." Cluster, Cloud and Grid Computing (CCGrid), 2014 14th IEEE/ACM International Symposium on. IEEE, 2014.

31. David A. Baum; Stacey D. Smith (2013). Tree Thinking: An Introduction to Phylogenetic Biology. Roberts. p. 442. ISBN 978-1-936221-16-5.

32. Felsenstein, Joseph (1985). "Confidence Limits on Phylogenies: An Approach Using the Bootstrap". Evolution. 39 (4): 783. doi:10.2307/2408678. ISSN 0014-3820.
33. Hillis, D. M.; Bull, J. J. (1993). "An Empirical Test of Bootstrapping as a Method for Assessing Confidence in Phylogenetic Analysis". Systematic Biology. 42 (2): 182–192. doi:10.1093/sysbio/42.2.182. ISSN 1063-5157.
34. Huelsenbeck, John; Rannala, Bruce (2004). "Frequentist Properties of Bayesian Posterior Probabilities of Phylogenetic Trees Under Simple and Complex Substitution Models". Systematic Biology. 53 (6): 904–913. doi:10.1080/10635150490522629. ISSN 1063-5157.
35. Chemisquy, M. A. A.; Prevosti, F. J. (2013). "Evaluating the clade size effect in alternative measures of branch support". Journal of Zoological Systematics and Evolutionary Research: n/a. doi:10.1111/jzs.12024.
36. Phillips, M. J.; Delsuc, F; Penny, D (2004). "Genome-Scale Phylogeny and the Detection of Systematic Biases". Molecular Biology and Evolution. 21 (7): 1455–8. doi:10.1093/molbev/msh137. PMID 15084674.
37. Goloboff, Pablo A.; Carpenter, James M.; Arias, J. Salvador; Esquivel, Daniel Rafael Miranda (2008). "Weighting against homoplasy improves phylogenetic analysis of morphological data sets". Cladistics. 24 (5): 758–773. doi:10.1111/j.1096-0031.2008.00209.x.
38. Goloboff, Pablo A. (1997). "Self-Weighted Optimization: Tree Searches and Character State Reconstructions under Implied Transformation Costs". Cladistics. 13 (3): 225–245. doi:10.1111/j.1096-0031.1997.tb00317.x.
39. Arnold, M.L. (1996). Natural Hybridization and Evolution. New York: Oxford University Press. p. 232. ISBN 978-0-19-509975-1.
40. Wendel, J F. & Doyle, J.J. (1998): DNA Sequencing. In Molecular Systematics of Plants II. Editors: D.E. Soltis, P.S. Soltis, J.J. Doyle. Kluwer, Boston, pp. 265–296.
41. Funk, D. J.; Omland, K. E. (2003). "Species-level paraphyly and polyphyly: Frequency, causes, and consequences, with insights from animal mitochondrial DNA". Annual Review of Ecology, Evolution, and Systematics. 34: 397–423. doi:10.1146/annurev.ecolsys.34.011802.132421.
42. "Genealogy of Life (GoLife)". National Science Foundation. Retrieved 5 May 2015. The GoLife program builds upon the AToL program by accommodating the complexity of diversification patterns across all of life's history. Our current knowledge of processes such as hybridization, endosymbiosis and lateral gene transfer makes clear that the evolutionary history of life on Earth cannot accurately be depicted as a single, typological, bifurcating tree.
43. Qu, Yanhua; Zhang, Ruiying; Quan, Qing; Song, Gang; Li, Shou Hsien; Lei, Fumin (December 2012). "Incomplete lineage sorting or secondary admixture: disentangling historical divergence from recent gene flow in the Vinous-throated parrotbill (Paradoxornis webbianus)". Molecular Ecology. 21 (24): 6117–6133. doi:10.1111/mec.12080.

44. Pollard, Daniel A.; Iyer, Venky N.; Moses, Alan M.; Eisen, Michael B. (2006). "Widespread Discordance of Gene Trees with Species Tree in Drosophila: Evidence for Incomplete Lineage Sorting". PLoS Genetics. 2 (10): e173. doi:10.1371/journal.pgen.0020173.

45. Zwickl, Derrick J.; Hillis, David M. (2002). "Increased Taxon Sampling Greatly Reduces Phylogenetic Error". Systematic Biology. 51 (4): 588–98. doi:10.1080/10635150290102339. PMID 12228001.

46. Wiens, John J. (2006). "Missing data and the design of phylogenetic analyses". Journal of Biomedical Informatics. 39 (1): 34–42. doi:10.1016/j.jbi.2005.04.001. PMID 15922672.

47. Blomberg, Simon P.; Garland Jr, Theodore; Ives, Anthony R. (2003). "Testing for phylogenetic signal in comparative data: Behavioral traits are more labile". Evolution. 57 (4): 717–45. doi:10.1111/j.0014-3820.2003.tb00285.x. PMID 12778543.

48. Archie, J.W. (1985). "Methods for coding variable morphological features for numerical taxonomic analysis". Systematic Zoology. 34 (3): 326–345. doi:10.2307/2413151.

49. Prevosti, Francisco J.; Chemisquy, María A. (2009). "The impact of missing data on real morphological phylogenies: Influence of the number and distribution of missing entries". Cladistics. 26 (3): 326–339. doi:10.1111/j.1096-0031.2009.00289.x.

50. Cobbett, Andrea; Wilkinson, Mark; Wills, Matthew (2007). "Fossils Impact as Hard as Living Taxa in Parsimony Analyses of Morphology". Systematic Biology. 56 (5): 753–66. doi:10.1080/10635150701627296. PMID 17886145.

51. Huelsenbeck, John P. (1994). "Comparing the Stratigraphic Record to Estimates of Phylogeny". Paleobiology. 20 (4): 470–83. JSTOR 2401230.

CHAPTER 4

Computational Neuroscience

Chapter Contents

Computational neuroscience is a diverse and interdisciplinary science. It combines many fields, such as cognitive science, electrical engineering, physics, and computer science into one cohesive field. Its aim is to explain a range of biological, brain, and neural system-related phenomena from a multidisciplinary standpoint. The information gained from such studies could be useful in the future of various medical fields. The overall, supreme goal of computational neuroscience is to explain the phenomenon of consciousness.

INTRODUCTION

Computational neuroscience (also theoretical neuroscience) studies brain function in terms of the information processing properties of the structures that make up the nervous system. It is an interdisciplinary computational science that links the diverse fields of neuroscience, cognitive science, and psychology with electrical engineering, computer science, mathematics, and physics.

Computational neuroscience is distinct from psychological connectionism and from learning theories of disciplines such as machine learning, neural networks, and computational learning theory in that it emphasizes descriptions of functional and biologically realistic neurons (and neural systems) and their physiology and dynamics. These models capture the essential features of the biological system at multiple spatial-temporal scales, from membrane currents, proteins, and chemical coupling to network oscillations, columnar and topographic architecture, and learning and memory.

These computational models are used to frame hypotheses that can be directly tested by biological or psychological experiments.

The term "computational neuroscience" was introduced by Eric L. Schwartz, who organized a conference, held in 1985 in Carmel, California, at the request of the Systems Development Foundation to provide a summary of the current status of a field which until that point was referred to by a variety of names, such as neural modeling, brain theory and neural networks. The proceedings of this definitional meeting were published in 1990 as the book *Computational Neuroscience*. The first open international meeting focused on Computational Neuroscience was organized by James M. Bower and John Miller in San Francisco, California in 1989 and has continued each year since as the annual CNS meeting. The first graduate educational program in computational neuroscience was organized as the Computational and Neural Systems Ph.D. program at the California Institute of Technology in 1985.

The early historical roots of the field can be traced to the work of people such as Louis Lapicque, Hodgkin & Huxley, Hubel & Wiesel, and David Marr, to name a few. Lapicque introduced the integrate and fire model of the neuron in a seminal article published in 1907; this model is still one of the most popular models in computational neuroscience for both cellular and neural networks studies, as well as in mathematical neuroscience because of its simplicity (see the recent review article for the centenary of Lapicque's original paper). About 40 years later, Hodgkin & Huxley developed the voltage clamp and created the first biophysical model of the action potential. Hubel & Wiesel discovered that neurons in the primary visual cortex, the first cortical area to process information coming from the retina, have oriented receptive fields and are organized in columns. David Marr's work focused on the interactions between neurons, suggesting computational approaches to the study of how functional groups of neurons within the hippocampus and neocortex interact, store, process, and transmit information. Computational modeling of biophysically realistic neurons and dendrites began with the work of Wilfrid Rall, with the first multicompartmental model using cable theory.

Major Topics

Research in computational neuroscience can be roughly categorized into several lines of inquiry. Most computational neuroscientists collaborate closely with experimentalists in analyzing novel data and synthesizing new models of biological phenomena.

Single-Neuron Modeling

Even single neurons have complex biophysical characteristics and can perform computations (e.g.). Hodgkin and Huxley's original model only employed two voltage-sensitive currents (Voltage sensitive ion channels are glycoprotein molecules which extend through the lipid bilayer, allowing ions to traverse under certain conditions through the axolemma), the fast-acting sodium and the inward-rectifying potassium. Though successful in predicting the timing and qualitative features of the action potential, it nevertheless failed to predict a number of important features such as adaptation and shunting. Scientists now believe that there are a wide variety of voltage-sensitive currents, and the implications of the differing dynamics, modulations, and sensitivity of these currents is an important topic of computational neuroscience.

The computational functions of complex dendrites are also under intense investigation. There is a large body of literature regarding how different currents interact with geometric properties of neurons.

Some models are also tracking biochemical pathways at very small scales such as spines or synaptic clefts.

There are many software packages, such as GENESIS and NEURON, that allow rapid and systematic *in silico* modeling of realistic neurons. Blue Brain, a project founded by Henry Markram from the École Polytechnique Fédérale de Lausanne, aims to construct a biophysically detailed simulation of a cortical column on the Blue Gene supercomputer.

A problem in the field is that detailed neuron descriptions are computationally expensive and this can handicap the pursuit of realistic network investigations, where many neurons need to be simulated. So, researchers that study large neural circuits typically represent each neuron and synapse simply, ignoring much of the biological detail. This is unfortunate as there is evidence that the richness of biophysical properties on the single neuron scale can supply mechanisms that serve as the building blocks for network dynamics. Hence there is a drive to produce simplified neuron models that can retain significant biological fidelity at a low computational overhead. Algorithms have been developed to produce faithful, faster running, simplified surrogate neuron models from computationally expensive, detailed neuron models.

Development, Axonal Patterning, and Guidance

How do axons and dendrites form during development? How do axons know where to target and how to reach these targets? How do neurons migrate to the proper position in the central and peripheral systems? How do synapses form? We know from molecular biology that distinct parts of the nervous system release distinct chemical cues, from growth factors to hormones that modulate and influence the growth and development of functional connections between neurons.

Theoretical investigations into the formation and patterning of synaptic connection and morphology are still nascent. One hypothesis that has recently garnered some attention is the *minimal wiring hypothesis*, which postulates that the formation of axons and dendrites effectively minimizes resource allocation while maintaining maximal information storage.

Sensory Processing

Early models of sensory processing understood within a theoretical framework are credited to Horace Barlow. Somewhat similar to the minimal wiring hypothesis described in the preceding section, Barlow understood the processing of the early sensory systems to be a form of efficient coding, where the neurons encoded information which minimized the number of spikes. Experimental and computational work have since supported this hypothesis in one form or another.

Current research in sensory processing is divided among a biophysical modelling of different subsystems and a more theoretical modelling of perception. Current models of perception have suggested that the brain performs some form of Bayesian inference and integration of different sensory information in generating our perception of the physical world.

Memory and Synaptic Plasticity

Earlier models of memory are primarily based on the postulates of Hebbian learning. Biologically relevant models such as Hopfield net have been developed to address the properties of associative, rather than content-addressable, style of memory that occur in biological systems. These attempts are primarily focusing on the formation of medium- and long-term memory, localizing in the hippocampus. Models of working memory, relying on theories of network oscillations and persistent activity, have been built to capture some features of the prefrontal cortex in context-related memory.

One of the major problems in neurophysiological memory is how it is maintained and changed through multiple time scales. Unstable synapses are easy to train but also prone to stochastic disruption. Stable synapses forget less easily, but they are also harder to consolidate. One recent computational hypothesis involves cascades of plasticity that allow synapses to function at multiple time scales. Stereochemically detailed models of the acetylcholine receptor-based synapse with the Monte Carlo method, working at the time scale of microseconds, have been built. It is likely that computational tools will contribute greatly to our understanding of how synapses function and change in relation to external stimulus in the coming decades.

Behaviors of Networks

Biological neurons are connected to each other in a complex, recurrent fashion. These connections are, unlike most artificial neural networks, sparse and usually specific. It is not known how information is transmitted through such sparsely connected networks. It is also unknown what the computational functions of these specific connectivity patterns are, if any.

The interactions of neurons in a small network can be often reduced to simple models such as the Ising model. The statistical mechanics of such simple systems are well-characterized theoretically. There has been some recent evidence that suggests that dynamics of arbitrary neuronal networks can be reduced to pairwise interactions. It is not known, however, whether such descriptive dynamics impart any important computational function. With the emergence of two-photon microscopy and calcium imaging, we now have powerful experimental methods with which to test the new theories regarding neuronal networks.

In some cases the complex interactions between *inhibitory* and *excitatory* neurons can be simplified using mean field theory, which gives rise to the population model of neural networks. While many neurotheorists prefer such models with reduced complexity, others argue that uncovering structural functional relations depends on including as much neuronal and network structure as possible. Models of this type are typically built in large simulation platforms like GENESIS or NEURON. There have been some attempts to provide unified methods that bridge and integrate these levels of complexity.

Cognition, Discrimination, and Learning

Computational modeling of higher cognitive functions has only recently begun. Experimental data comes primarily from single-unit recording in primates. The frontal lobe and parietal lobe function as integrators of information from multiple sensory modalities. There are some tentative ideas regarding how simple mutually inhibitory functional circuits in these areas may carry out biologically relevant computation.

The brain seems to be able to discriminate and adapt particularly well in certain contexts. For instance, human beings seem to have an enormous capacity for memorizing and recognizing faces. One of the key goals of computational neuroscience is to dissect how biological systems carry out these complex computations efficiently and potentially replicate these processes in building intelligent machines.

The brain's large-scale organizational principles are illuminated by many fields, including biology, psychology, and clinical practice. Integrative neuroscience attempts to consolidate these observations through unified descriptive models and databases of behavioral measures and recordings. These are the bases for some quantitative modeling of large-scale brain activity.

The Computational Representational Understanding of Mind (CRUM) is another attempt at modeling human cognition through simulated processes like acquired rule-based systems in decision making and the manipulation of visual representations in decision making.

Consciousness

One of the ultimate goals of psychology/neuroscience is to be able to explain the everyday experience of conscious life. Francis Crick and Christof Koch made some attempts to formulate a consistent framework for future work in neural correlates of consciousness (NCC), though much of the work in this field remains speculative.

Computational Clinical Neuroscience

It is a field that brings together experts in neuroscience, neurology, psychiatry, decision sciences and computational modeling to quantitatively define and investigate problems in neurological and psychiatric diseases, and to train scientists and clinicians that wish to apply these models to diagnosis and treatment.

Notable Persons

- Phil Husbands, professor of computer science and artificial intelligence at the English University of Sussex
- Read Montague, American neuroscientist and popular science author
- Tomaso Poggio, Eugene McDermott professor in the Department of Brain and Cognitive Sciences, investigator at the McGovern Institute for Brain Research, a member of the MIT Computer Science and Artificial Intelligence Laboratory (CSAIL) and director of both the Center for Biological and Computational Learning at MIT and the Center for Brains, Minds, and Machines.
- Terry Sejnowski, investigator at the Howard Hughes Medical Institute and the Francis Crick Professor at The Salk Institute for Biological Studies where he directs the Computational Neurobiology Laboratory
- Haim Sompolinsky, William N. Skirball Professor of Neuroscience at the Edmond and Lily Safra Center for Brain Sciences (formerly the Interdisciplinary Center for Neural Computation), and a Professor of Physics at the Racah Institute of Physics at The Hebrew University of Jerusalem, Israel.

BIOLOGICAL NEURON MODEL

A **biological neuron model**, also known as a **spiking neuron model**, is a mathematical description of the properties of certain cells in the nervous system that generate sharp electrical potentials across their cell membrane, roughly one millisecond in duration, as shown in Fig. 1. Spiking neurons are known to be a major signaling unit of the nervous system, and for this reason characterizing their operation is of great importance. It is worth noting that not all the cells of the nervous system produce the type of spike that define the scope of the spiking neuron models. For example, cochlear hair cells, retinal receptor cells, and retinal bipolar cells do not spike. Furthermore, many cells in the nervous system are not classified as neurons but instead are classified as glia.

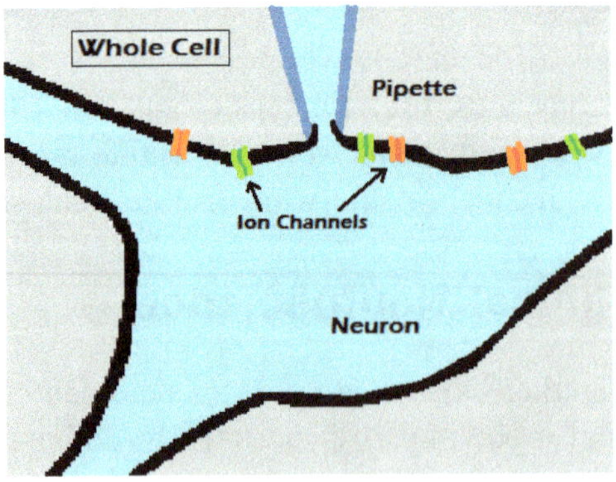

"Whole cell" measurement technique, which captures the spiking activity of a single neuron and produces full amplitude action potentials.

Ultimately, biological neuron models aim to explain the mechanisms underlying the operation of the nervous system for the purpose of restoring lost control capabilities such as perception (e.g. deafness or blindness), motor movement decision making, and continuous limb control. In that sense, biological neural models differ from artificial neuron models that do not presume to predict the outcomes of experiments involving the biological neural tissue (although artificial neuron models are also concerned with execution of perception and estimation tasks). Accordingly, an important aspect of biological neuron models is experimental validation, and the use of physical units to describe the experimental procedure associated with the model predictions.

Neuron models can be divided into two categories according to the physical units of the interface of the model. Each category could be further divided according to the abstraction/detail level:

- Electrical input–output membrane voltage models – These models produce a prediction for membrane output voltage as function of electrical stimulation at the input stage (either voltage or current). The various models in this category differ in the exact functional relationship between the input current and the output voltage and in the level of details. Some models in this category are black box models and distinguish only between two measured voltage levels: the presence of a spike (also known as "action potential") or a quiescent state. Other models are more detailed and account for sub-cellular processes.

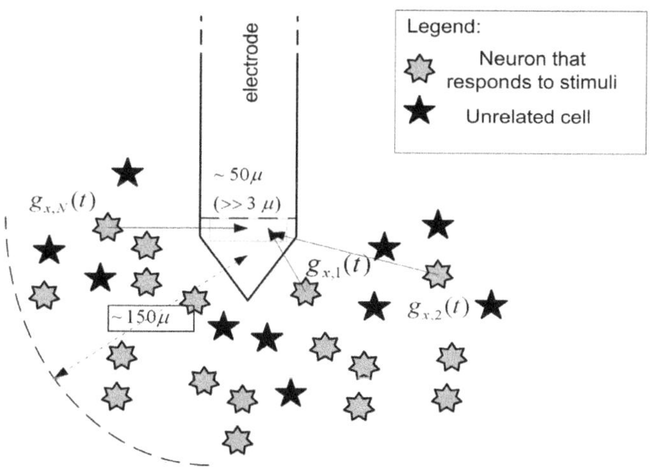

Extracellular measurement: Captures spikes with lower amplitudes, often from several spiking sources, depending on the size of the electrode and its proximity to the sources. Despite the decreased amplitude levels produced by this technique, it also has several advantages: 1) Is easier to obtain experimentally; 2) Is robust and lasts for a longer time; 3) Can reflect the dominant effect, especially when conducted in an anatomical region with many similar cells.

- Natural or pharmacological input neuron models – The models in this category connect between the input stimulus which can be either pharmacological or natural, to the probability of a spike event. The input stage of these models is not electrical, but rather has either pharmacological (chemical) concentration units, or physical units that characterize an external stimulus such as light, sound or other forms of physical pressure. Furthermore, the output stage represents the probability of a spike event and not an electrical voltage. Typically, this output probability is normalized (divided by) a time constant, and the resulting normalized probability is called the "firing rate" and has units of Hertz. The probabilistic description taken by the models in this category was inspired from laboratory experiments involving either natural or pharmacological stimulation which exhibit variability in the resulting spike pattern. Nevertheless, when averaging these experimental results across several trials, a clear pattern is often revealed.

Although it is not unusual in science and engineering to have several descriptive models for different abstraction/detail levels, the number of different, sometimes contradicting, biological neuron models is exceptionally high. This situation is partly the result of the many different experimental settings, and the difficulty to separate the intrinsic properties of a single neuron from measurements effects and interactions of many cells (network effects). To accelerate the convergence to a unified theory, we list several models in each category, and where applicable, also references to supporting experiments.

Electrical Input–Output Membrane Voltage Models

The models in this category describe the relationship between neuronal membrane currents at the input stage, and membrane voltage at the output stage. The most extensive experimental inquiry in this category of models was made by Hodgkin–Huxley in the early 1950s using an experimental setup that punctured the cell membrane and allowed to force a specific membrane voltage/current.

Most modern electrical neural interfaces apply extra-cellular electrical stimulation to avoid membrane puncturing which can lead to cell death and tissue damage. Hence, it is not clear to what extent the electrical neuron models hold for extra-cellular stimulation.

Integrate-and-Fire

One of the earliest models of a neuron was first investigated in 1907 by Louis Lapicque. A neuron is represented in time by

$$I(t) = C_{\mathrm{m}} \frac{dV_{\mathrm{m}}(t)}{dt}$$

which is just the time derivative of the law of capacitance, $Q = CV$. When an input current is applied, the membrane voltage increases with time until it reaches a constant threshold V_{th}, at which point a delta function spike occurs and the voltage is reset to its resting potential, after which the model continues to run. The *firing frequency* of the model thus increases linearly without bound as input current increases.

The model can be made more accurate by introducing a refractory period t_{ref} that limits the firing frequency of a neuron by preventing it from firing during that period. Through some calculus involving a Fourier transform, the firing frequency as a function of a constant input current thus looks like

$$f(I) = \frac{I}{C_{\mathrm{m}} V_{\mathrm{th}} + t_{\mathrm{ref}} I}.$$

A remaining shortcoming of this model is that it implements no time-dependent memory. If the model receives a below-threshold signal at some time, it will retain that voltage boost forever until it fires again. This characteristic is clearly not in line with observed neuronal behavior.

Hodgkin–Huxley Model

The Hodgkin–Huxley model (H&H model) is a model of the relationship between ion currents crossing the neuronal cell membrane and the membrane voltage. The model is based on experiments that allowed to force membrane voltage using an intra-cellular pipette. This model is based on the concept of membrane ion channels and relies on data from the squid giant axon. Hodgkin-Huxley was awarded the 1963 Nobel Prize in Physiology or Medicine for this model.

We note as before our voltage-current relationship, this time generalized to include multiple voltage-dependent currents:

$$C_{\mathrm{m}} \frac{dV(t)}{dt} = -\sum_{i} I_i(t, V).$$

Each current is given by Ohm's Law as

$$I(t, V) = g(t, V) \cdot (V - V_{\mathrm{eq}})$$

where $g(t,V)$ is the conductance, or inverse resistance, which can be expanded in terms of its constant average and the activation and inactivation fractions m and h, respectively, that determine how many ions can flow through available membrane channels. This expansion is given by

$$g(t, V) = \bar{g} \cdot m(t, V)^p \cdot h(t, V)^q$$

and our fractions follow the first-order kinetics

$$\frac{dm(t, V)}{dt} = \frac{m_{\infty}(V) - m(t, V)}{\tau_{\mathrm{m}}(V)} = \alpha_{\mathrm{m}}(V) \cdot (1 - m) - \beta_{\mathrm{m}}(V) \cdot m$$

with similar dynamics for h, where we can use either τ and m_{∞} or α and β to define our gate fractions.

With such a form, all that remains is to individually investigate each current one wants to include. Typically, these include inward Ca^{2+} and Na^{+} input currents and several varieties of K^{+}outward currents, including a "leak" current.

The end result can be at the small end 20 parameters which one must estimate or measure for an accurate model, and for complex systems of neurons not easily tractable by computer. Careful simplifications of the Hodgkin–Huxley model are therefore needed.

Leaky Integrate-and-Fire

In the leaky integrate-and-fire model, the memory problem is solved by adding a "leak" term to the membrane potential, reflecting the diffusion of ions that occurs through the membrane when some equilibrium is not reached in the cell. The model looks like

$$I(t) - \frac{V_{\rm m}(t)}{R_{\rm m}} = C_{\rm m} \frac{dV_{\rm m}(t)}{dt}$$

where R_m is the membrane resistance, as we find it is not a perfect insulator as assumed previously. This forces the input current to exceed some threshold $I_{th} = V_{th} / R_m$ in order to cause the cell to fire, else it will simply leak out any change in potential. The firing frequency thus looks like

$$f(I) = \begin{cases} 0, & I \le I_{\rm th} \\ [t_{\rm ref} - R_{\rm m} C_{\rm m} \log(1 - \frac{V_{\rm th}}{I R_{\rm m}})]^{-1}, & I > I_{\rm th} \end{cases}$$

which converges for large input currents to the previous leak-free model with refractory period.

Fractional-Order Leaky Integrate-and-Fire

Recent advances in computational and theoretical fractional calculus lead to a new form of model, called Fractional-Order Leaky integrate-and-fire developed by Teka et al. The great advantage of this model is that it can capture and integrate all the past activities and can reproduce the time dependent spiking adaptations observed on pyramidal neurons. The model has the following form more details can be found in

$$I(t) - \frac{V_{\rm m}(t)}{R_{\rm m}} = C_{\rm m} \frac{d^{\alpha} V_{\rm m}(t)}{d^{\alpha} t}$$

Galves-Löcherbach

3D visualization of the Galves-Löcherbach model for biological neural nets. This visualization is set for 4,000 neurons (4 layers with one population of inhibitory neurons and one population of excitatory neurons each) at 180 intervals of time.

The Galves-Löcherbach model is a specific development of the leaky integrate-and-fire model. It is inherently stochastic. It was developed by mathematicians Antonio Galves and Eva Löcherbach. Given the model specifications, the probability that a given i may be described by

$$\mathrm{Prob}(X_t(i) = 1 | \mathcal{F}_{t-1}) = \phi_i \left(\sum_{j \in I} W_{j \to i} \sum_{s=L_t^i}^{t-1} g_j(t-s) X_s(j), t - L_t^i \right),$$

where $W_{j \to i}$ is a synaptic weight, describing the influence of neuron j on neuron i, g_j xpresses the leak, andL_t^i provides the spiking history of neuron *i* before *t*, according to $L_t^i = \sup\{s < t : X_s(i) = 1\}$.

Exponential Integrate-and-Fire

In the Exponential Integrate-and-Fire, spike generation is exponential, following the equation:

$$\frac{dX}{dt} = \Delta_T \exp\left(\frac{X - X_T}{\Delta_T} \right).$$

where X is the membrane potential, X_T is the membrane potential threshold, and Δ_T is the sharpness of action potential initiation, usually around 1 mV for cortical pyramidal neurons. Once the membrane potential crosses X_T, it diverges to infinity in finite time.

FitzHugh–Nagumo

Sweeping simplifications to Hodgkin–Huxley were introduced by FitzHugh and Nagumo in 1961 and 1962. Seeking to describe "regenerative self-excitation" by a nonlinear positive-feedback membrane voltage and recovery by a linear negative-feedback gate voltage, they developed the model described by

$$\frac{dV}{dt} = V - V^3 - w + I_{\mathrm{ext}}$$

$$\tau \frac{dw}{dt} = V - a - bw$$

where we again have a membrane-like voltage and input current with a slower general gate voltage *w* and experimentally-determined parameters $a = -0.7$, $b = 0.8$, $\tau = 1/0.08$. Although not clearly derivable from biology, the model allows for a simplified, immediately available dynamic, without being a trivial simplification.

Morris–Lecar

In 1981 Morris and Lecar combined Hodgkin–Huxley and FitzHugh–Nagumo into a voltage-gated calcium channel model with a delayed-rectifier potassium channel, represented by

$$C\frac{dV}{dt} = -I_{\text{ion}}(V,w) + I$$

$$\frac{dw}{dt} = \phi \cdot \frac{w_\infty - w}{\tau_w}$$

where $I_{\text{ion}}(V,w) = \bar{g}_{\text{Ca}} m_\infty \cdot (V - V_{\text{Ca}}) + \bar{g}_{\text{K}} w \cdot (V - V_{\text{K}}) + \bar{g}_{\text{L}} \cdot (V - V_{\text{L}})$.

Hindmarsh–Rose

Building upon the FitzHugh–Nagumo model, Hindmarsh and Rose proposed in 1984 a model of neuronal activity described by three coupled first order differential equations:

$$\frac{dx}{dt} = y + 3x^2 - x^3 - z + I$$

$$\frac{dy}{dt} = 1 - 5x^2 - y$$

$$\frac{dz}{dt} = r \cdot (4(x + \tfrac{8}{5}) - z)$$

with $r^2 = x^2 + y^2 + z^2$, and $r \approx 10^{-2}$ so that the z variable only changes very slowly. This extra mathematical complexity allows a great variety of dynamic behaviors for the membrane potential, described by the x variable of the model, which include chaotic dynamics. This makes the Hindmarsh–Rose neuron model very useful, because being still simple, allows a good qualitative description of the many different patterns of the action potential observed in experiments.

Cable Theory

Cable theory describes the dendritic arbor as a cylindrical structure undergoing a regular pattern of bifurcation, like branches in a tree. For a single cylinder or an entire tree, the input conductance at the base (where the tree meets the cell body, or any such boundary) is defined as

$$G_{in} = \frac{G_\infty \tanh(L) + G_L}{1 + (G_L/G_\infty)\tanh(L)},$$

where L is the electrotonic length of the cylinder which depends on its length, diameter, and resistance. A simple recursive algorithm scales linearly with the number of branches and can be used to calculate the effective conductance of the tree. This is given by

$$G_D = G_m A_D \tanh(L_D)/L_D$$

where $A_D = \pi ld$ is the total surface area of the tree of total length l, and L_D is its total electrotonic length. For an entire neuron in which the cell body conductance is G_S and the membrane conductance per unit area is $G_{md} = G_m / A$, we find the total neuron conductance G_N for n dendrite trees by adding up all tree and soma conductances, given by

$$G_N = G_S + \sum_{j=1}^{n} A_{D_j} F_{dga_j},$$

where we can find the general correction factor F_{dga} experimentally by noting $G_D = G_{md} A_D F_{dga}$.

Compartmental Models

The cable model makes a number of simplifications to give closed analytic results, namely that the dendritic arbor must branch in diminishing pairs in a fixed pattern. A compartmental model allows for any desired tree topology with arbitrary branches and lengths, but makes simplifications in the interactions between branches to compensate. Thus, the two models give complementary results, neither of which is necessarily more accurate.

Each individual piece, or compartment, of a dendrite is modeled by a straight cylinder of arbitrary length l and diameter d which connects with fixed resistance to any number of branching cylinders. We define the conductance ratio of the ith cylinder as $B_i = G_i / G_\infty$, where

$$G_\infty = \frac{\pi d^{3/2}}{2\sqrt{R_i R_m}}$$

and R_i is the resistance between the current compartment and the next. We obtain a series of equations for conductance ratios in and out of a compartment by making corrections to the normal dynamic $B_{out,i} = B_{in,i+1}$, as

- $B_{\mathrm{out},i} = \dfrac{B_{\mathrm{in},i+1}(d_{i+1}/d_i)^{3/2}}{\sqrt{R_{\mathrm{m},i+1}/R_{\mathrm{m},i}}}$
- $B_{\mathrm{in},i} = \dfrac{B_{\mathrm{out},i} + \tanh X_i}{1 + B_{\mathrm{out},i} \tanh X_i}$
- $B_{\mathrm{out,par}} = \dfrac{B_{\mathrm{in,dau1}}(d_{\mathrm{dau1}}/d_{\mathrm{par}})^{3/2}}{\sqrt{R_{\mathrm{m,dau1}}/R_{\mathrm{m,par}}}} + \dfrac{B_{\mathrm{in,dau2}}(d_{\mathrm{dau2}}/d_{\mathrm{par}})^{3/2}}{\sqrt{R_{\mathrm{m,dau2}}/R_{\mathrm{m,par}}}} + \ldots$

where the last equation deals with parents and daughters at branches, and $X_i = \frac{l_i\sqrt{4R_i}}{\sqrt{d_i R_m}}$. We can iterate these equations through the tree until we get the point where the dendrites connect to the cell body (soma), where the conductance ratio is $B_{in,stem}$. Then our total neuron conductance is given by

$$G_N = \frac{A_{\mathrm{soma}}}{R_{\mathrm{m,soma}}} + \sum_j B_{\mathrm{in,stem},j} G_{\infty,j}.$$

An example of a compartmental model of a neuron, with an algorithm to reduce the number of compartments (increase the computational speed) and yet retain the salient electrical characteristics, can be found in.

Natural Input Stimulus Neuron Models

The models in this category were derived following experiments involving natural stimulation such as light, sound, touch, or odor. In these experiments, the spike pattern resulting from each stimulus presentation varies from trial to trial, but the averaged response from several trials often converges to a clear pattern. Consequently, the models in this category generate a probabilistic relationship between the input stimulus to spike occurrences.

The Non-Homogeneous Poisson Process Model (Siebert)

Siebert modeled the neuron spike firing pattern using a non-homogeneous Poisson process model, following experiments involving the auditory system. According to Siebert, the probability of a spiking event at the time interval $[t, t + \Delta_t]$ is proportional to a non negative function $g[s(t)]$, where s(t) is the raw stimulus.:

$$P_{spike}(t \in [t', t' + \Delta_t]) = \Delta_t \cdot g[s(t)]$$

Siebert considered several functions as $g[s(t)]$, including $g[s(t)] \propto s^2(t)$ for low stimulus intensities.

The main advantage of Siebert's model is its simplicity. The shortcomings of the model is its inability to reflect properly the following phenomena:

- The edge emphasizing property of the neuron in response to a stimulus pulse.
- The saturation of the firing rate.
- The values of inter-spike-interval-**histogram** at short intervals values (close to zero).

These shortcoming are addressed by the two state Markov Model.

The two State Markov Model (Nossenson & Messer)

The spiking neuron model by Nossenson & Messer produces the probability of the neuron to fire a spike as a function of either an external or pharmacological stimulus.The model consists of a cascade of a receptor layer model and a spiking neuron model, as shown in Fig 4. The connection between the external stimulus to the spiking probability is made in two steps: First, a receptor cell model translates the raw external stimulus to neurotransmitter concentration, then, a spiking neuron model connects between neurotransmitter concentration to the firing rate (spiking probability). Thus, the spiking neuron model by itself depends on neurotransmitter concentration at the input stage.

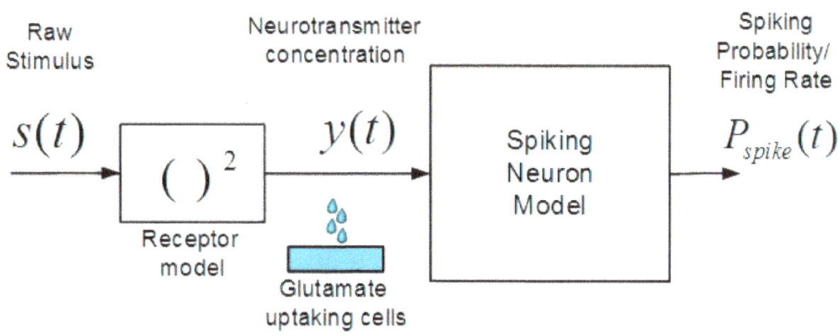

High level block diagram of the receptor layer and neuron model by Nossenson & Messer.

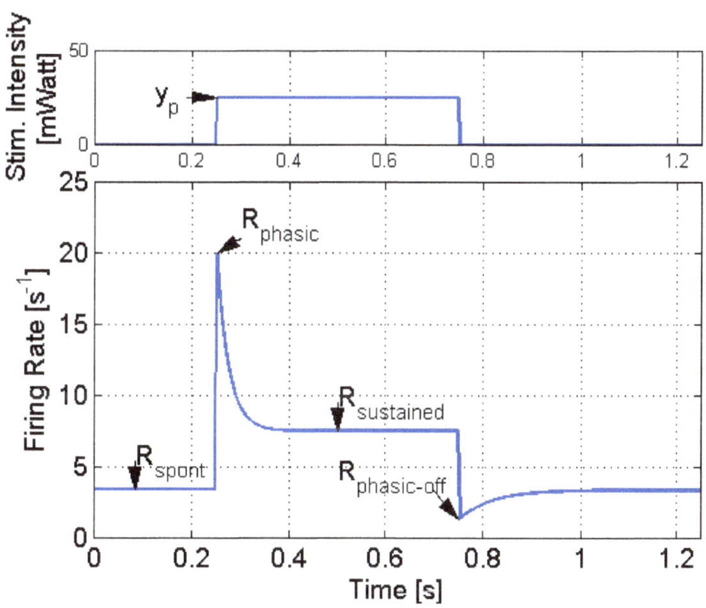

The prediction for the firing rate in response to a pulse stimulus as given by the model by Nossenson & Messer.

An important feature of this model is the prediction for neurons firing rate pattern which captures, using a low number of free parameters, the characteristic edge emphasized response of neurons to a stimulus pulse, as shown in Fig. 5. The firing rate is identified both as a normalized probability for neural spike firing, and as a quantity proportional to the current of neurotransmitters released by the cell. The expression for the firing rate takes the following form:

$$R_{fire}(t) = \frac{P_{spike}(t;\Delta_t)}{\Delta_t} = [y(t) + R_0] \cdot P_0(t)$$

where,

- P0 is the probability of the neuron to be "armed" and ready to fire. It is given by the following differential equation:

$$\dot{P}_0 = -[y(t) + R_0 + R_1] \cdot P_0(t) + R_1$$

P0 could be generally calculated recursively using Euler method, but in the case of a pulse of stimulus it yields a simple closed form expression.

- y(t) is the input of the model and is interpreted as the neurotransmitter concentration on the cell surrounding (in most cases glutamate) . For an external stimulus it can be estimated through the receptor layer model:

$y(t) \simeq g_{gain} \cdot \langle s^2(t) \rangle$, with $\langle s^2(t) \rangle$ being short temporal average of stimulus power (given in Watt or other energy per time unit).

- R0 corresponds to the intrinsic spontaneous firing rate of the neuron.
- R1 is the recovery rate of the neuron from refractory state.

Other predictions by this model include:

1) The averaged Evoked Response Potential (ERP) due to population of many neurons in unfiltered measurements resembles the firing rate.
2) The voltage variance of activity due to multiple neuron activity resembles the firing rate (also known as Multi-Unit-Activity power or MUA).
3) The inter-spike-interval probability distribution takes the form a gamma-distribution like function.

Experimental evidence supporting the model by Nossenson & Messer

Property of the Model by Nossenson & Messer	References	Description of experimental evidence
The shape of the firing rate in response to an auditory stimulus pulse	[19][20][21][22][23]	The Firing Rate has the same shape of Fig 5.
The shape of the firing rate in response to a visual stimulus pulse	[24][25][26][27]	The Firing Rate has the same shape of Fig 5.
The shape of the firing rate in response to an olfactory stimulus pulse	[28]	The Firing Rate has the same shape of Fig 5.
The shape of the firing rate in response to a somato-sensory stimulus	[29]	The Firing Rate has the same shape of Fig 5.
The change in firing rate in response to neurotransmitter application (mostly glutamate)	[30][31]	Firing Rate change in response to neurotransmitter application (Glutamate)

Experimental evidence supporting the model by Nossenson & Messer

Property of the Model by Nossenson & Messer	References	Description of experimental evidence
Square dependence between an auditory stimulus pressure and the firing rate	[32]	Square Dependence between Auditory Stimulus pressure and the Firing Rate (- Linear dependence in pressure square (power)).
Square dependence between visual stimulus electric field (volts) and the firing rate	[25]	Square dependence between visual stimulus electric field (volts) - Linear Dependence between Visual Stimulus *Power* and the Firing Rate.
The shape of the Inter-Spike-Interval Statistics (ISI)	[33]	ISI shape resembles the gamma-function-like
The ERP resembles the firing rate in unfiltered measurements	[34]	The shape of the averaged evoked response potential in response to stimulus resembles the firing rate (Fig. 5).
MUA power resembles the firing rate	[17][35]	The shape of the empirical variance of extra-cellular measurements in response to stimulus pulse resembles the firing rate (Fig. 5).

Non-Markovian Models

The following is a list of published non-Markovian neuron models:

- Johnson, and Swami
- Berry and Meister
- Kass and Ventura

Pharmacological Input Stimulus Neuron Models

The models in this category produce predictions for experiments involving pharmacological stimulation.

Synaptic transmission (Koch & Segev)

According to the model by Koch and Segev, the response of a neuron to individual neurotransmitters can be modeled as an extension of the classical Hodgkin–Huxley model with both standard and nonstandard kinetic currents. Four neurotransmitters primarily have influence in the CNS. AMPA/kainate receptors are fast excitatory mediators while NMDA receptors mediate considerably slower currents. Fast inhibitory currents go through $GABA_A$ receptors, while $GABA_B$ receptors mediate by secondary *G*-protein-activated potassium channels. This range of mediation produces the following current dynamics:

- $I_{\text{AMPA}}(t,V) = \bar{g}_{\text{AMPA}} \cdot [O] \cdot (V(t) - E_{\text{AMPA}})$
- $I_{\text{NMDA}}(t,V) = \bar{g}_{\text{NMDA}} \cdot B(V) \cdot [O] \cdot (V(t) - E_{\text{NMDA}})$
- $I_{\text{GABA}_\text{A}}(t,V) = \bar{g}_{\text{GABA}_\text{A}} \cdot ([O_1] + [O_2]) \cdot (V(t) - E_{\text{Cl}})$
- $I_{\text{GABA}_\text{B}}(t,V) = \bar{g}_{\text{GABA}_\text{B}} \cdot \frac{[G]^n}{[G]^n + K_\text{d}} \cdot (V(t) - E_\text{K})$

where is the maximal conductance (around 1S) and E is the equilibrium potential of the given ion or transmitter (AMDA, NMDA, Cl, or K), while O describes the fraction of receptors that are open. For NMDA, there is a significant effect of *magnesium block* that depends sigmoidally on the concentration of intracellular magnesium by $B(V)$. For GABA_B, G is the concentration of the G-protein, and K_d describes the dissociation of G in binding to the potassium gates.

The dynamics of this more complicated model have been well-studied experimentally and produce important results in terms of very quick synaptic potentiation and depression, that is, fast, short-term learning.

The Two State Markov Model (Nossenson & Messer)

The model by Nossenson and Messer translates neurotransmitter concentration at the input stage to the probability of releasing neurotransmitter at the output stage. For a more detailed description of this model, see the Two state Markov model section above.

Conjecture 1: Relation between Artificial and Biological Neuron Models

The most basic model of a neuron consists of an input with some synaptic weight vector and an activation function or transfer function inside the neuron determining output. This is the basic structure used in artificial neurons, which in a neural network often looks like

$$y_i = \phi\left(\sum_j w_{ij} x_j\right)$$

where y_i is the output of the i th neuron, x_j is the jth input neuron signal, w_{ij} is the synaptic weight (or strength of connection) between the neurons i and j, and is the activation function. While this model has seen success in machine-learning applications, it is a poor model for real (biological) neurons, because it lacks the time-dependence that real neurons exhibit. Some of the earliest biological models took this form until kinetic models such as the Hodgkin–Huxley model became dominant.

In the case of modelling a biological neuron, physical analogues are used in place of abstractions such as "weight" and "transfer function". A neuron is filled and surrounded with water containing ions, which carry electric charge. The neuron is bound by an insulating cell membrane and can maintain a concentration of charged ions on either side that determines a capacitance C_m. The firing of a neuron involves the movement of ions into the cell that occurs when neurotransmitters cause ion channels on the cell membrane to open. We describe this by a physical time-dependent current $I(t)$. With this comes a change in voltage, or the electrical potential energy difference between the cell and its surroundings, which is observed to sometimes result in a voltage spike called an action potential which travels the length of the cell and triggers the release of further neurotransmitters. The voltage, then, is the quantity of interest and is given by $V_m(t)$.

Conjecture 3: The Neurotransmitter Based Energy Detection Scheme

The neurotransmitter based energy detection scheme suggests that the neural tissue chemically executes a Radar-like detection procedure.

As shown in Fig. 6, the key idea of the conjecture is to account neurotransmitter concentration, neurotransmitter generation and neurotransmitter removal rates as the important quantities in executing the detection task, while referring to the measured electrical potentials as a side effect that only in certain conditions coincide with the functional purpose of each step. The detection scheme is similar to a radar like "energy detection" because it includes signal squaring, temporal summation and a threshold switch mechanism, just like the energy detector, but it also includes a unit that emphasizes stimulus edges and a variable memory length (variable memory). According to this conjecture, the physiological equivalent of the energy test statistics is neurotransmitter concentration, and the firing rate corresponds to neurotransmitter current. The advantage of this interpretation is that it leads to a unit consistent explanation which allows to bridge between electrophysiological measurements, biochemical measurements and psychophysical results.

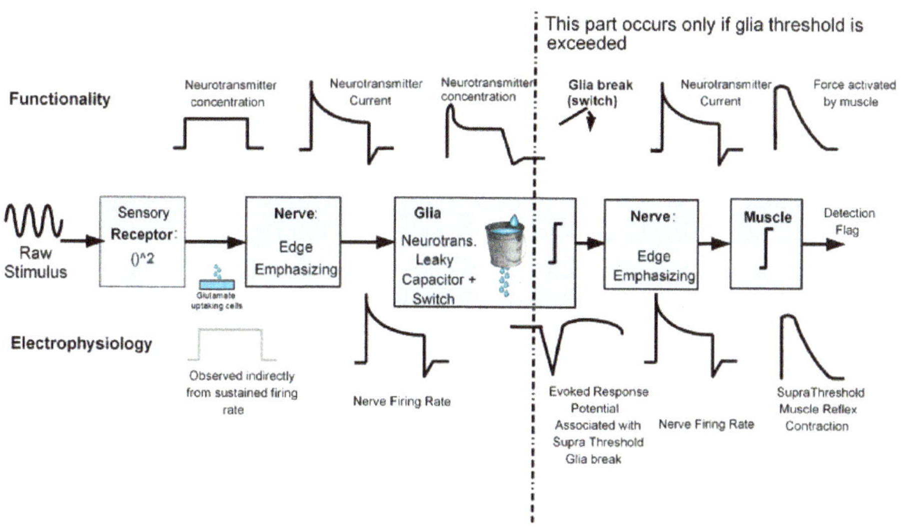

The biological neural detection scheme as suggested by Nossenson et al

The evidence reviewed in suggest the following association between functionality to histological classification:

- Stimulus squaring is likely to be perform by receptor cells.
- Stimulus edge emphasizing and signal transduction is performed by neurons.
- Temporal accumulation (=memory) is performed by glia cells. Short term neurotransmitter accumulation is likely to occur also in some types of neurons.
- Logical switching is executed by glia cells, and it results from exceeding a threshold level of neurotransmitter concentration. This threshold crossing is also accompanied by a stop in the growth of neurotransmitter leak rate.
- Physical all-or-non movement switching is due to muscle cells and results from exceeding a certain neurotransmitter concentration threshold on muscle surroundings.

Note that although the electrophysiological signals in Fig.6 are often similar to the functional signal (signal power / neurotransmitter concentration / muscle force), there are some stages in which the electrical observation is different from the functional purpose of the corresponding step. In particular, Nossenson et al suggested that glia threshold crossing has a completely different functional operation compared to the radiated electrophysiological signal, and that the latter might only be a side effect of glia break.

General Comments Regarding the Modern Perspective of Scientific and Engineering Models

- The models above are still idealizations. Corrections must be made for the increased membrane surface area given by numerous dendritic spines, temperatures significantly hotter than room-temperature experimental data, and nonuniformity in the cell's internal structure. Certain observed effects do not fit into some of these models. For instance, the temperature cycling (with minimal net temperature increase) of the cell membrane during action potential propagation not compatible with models which rely on modeling the membrane as a resistance which must dissipate energy when current flows through it. The transient thickening of the cell membrane during action potential propagation is also not predicted by these models, nor is the changing capacitance and voltage spike that results from this thickening incorporated into these models. The action of some anesthetics such as inert gases is problematic for these models as well. New models, such as the soliton model attempt to explain these phenomena, but are less developed than older models and have yet to be widely applied. Also improbable possibility of modelling of local chronobiology mechanisms.
- Modern views regarding of the role of the scientific model suggest that "All models are wrong but some are useful" (Box and Draper, 1987, Gribbin, 2009; Paninski et al., 2009).

BAYESIAN APPROACHES TO BRAIN FUNCTION

Bayesian approaches to brain function investigate the capacity of the nervous system to operate in situations of uncertainty in a fashion that is close to the optimal prescribed by Bayesian statistics. This term is used in behavioural sciences and neuroscience and studies associated with this term often strive to explain the brain's cognitive abilities based on statistical principles. It is frequently assumed that the nervous system maintains internal probabilistic models that are updated by neural processing of sensory information using methods approximating those of Bayesian probability.

Origins

This field of study has its historical roots in numerous disciplines including machine learning, experimental psychology and Bayesian statistics. As early as the 1860s, with the work of Hermann Helmholtz in experimental psychology the brain's ability to extract perceptual information from sensory data was modeled in terms of probabilistic estimation. The basic idea is that the nervous system needs to organize sensory data into an accurate internal model of the outside world.

Bayesian probability has been developed by many important contributors. Pierre-Simon Laplace, Thomas Bayes, Harold Jeffreys, Richard Cox and Edwin Jaynes developed mathematical techniques and procedures for treating probability as the degree of plausibility that could be assigned to a given supposition or hypothesis based on the available evidence. In 1988 E.T. Jaynes presented a framework for using Bayesian Probability to model mental processes. It was thus realized early on that the Bayesian statistical framework holds the potential to lead to insights into the function of the nervous system.

This idea was taken up in research on unsupervised learning, in particular the Analysis by Synthesis approach, branches of machine learning. In 1983 Geoffrey Hinton and colleagues proposed the brain could be seen as a machine making decisions based on the uncertainties of the outside world. During the 1990s researchers including Peter Dayan, Geoffrey Hinton and Richard Zemel proposed that the brain represents knowledge of the world in terms of probabilities and made specific proposals for tractable neural processes that could manifest such a Helmholtz Machine.

Psychophysics

A wide range of studies interpret the results of psychophysical experiments in light of Bayesian perceptual models. Many aspects of human perceptual and motor behavior can be modeled with Bayesian statistics. This approach, with its emphasis on behavioral outcomes as the ultimate expressions of neural information processing, is also known for modeling sensory and motor decisions using Bayesian decision theory. Examples are the work of Landy, Jacobs, Jordan, Knill, Kording and Wolpert, and Goldreich.

Neural Coding

Many theoretical studies ask how the nervous system could implement Bayesian algorithms. Examples are the work of Pouget, Zemel, Deneve, Latham, Hinton and Dayan. George and Hawkins published a paper that establishes a model of cortical information processing called hierarchical temporal memory that is based on Bayesian network of Markov chains. They further map this mathematical model to the existing knowledge about the architecture of cortex and show how neurons could recognize patterns by hierarchical Bayesian inference.

Electrophysiology

A number of recent electrophysiological studies focus on the representation of probabilities in the nervous system. Examples are the work of Shadlen and Schultz.

Predictive Coding

Predictive coding is a neurobiologically plausible scheme for inferring the causes of sensory input based on minimizing prediction error. These schemes are related formally to Kalman filtering and other Bayesian update schemes.

Free Energy

During the 1990s some researchers such as Geoffrey Hinton and Karl Friston began examining the concept of free energy as a calculably tractable measure of the discrepancy between actual features of the world and representations of those features captured by neural network models. A synthesis has been attempted recently by Karl Friston, in which the Bayesian brain emerges from a general principle of free energy minimisation. In this framework, both action and perception are seen as a consequence of suppressing free-energy, leading to perceptual and active inference and a more embodied (enactive) view of the Bayesian brain. Using variational Bayesian methods, it can be shown how internal models of the world are updated by sensory information to minimize free energy or the discrepancy between sensory input and predictions of that input. This can be cast (in neurobiologically plausible terms) as predictive coding or, more generally, Bayesian filtering.

According to Friston:

"The free-energy considered here represents a bound on the surprise inherent in any exchange with the environment, under expectations encoded by its state or configuration. A system can minimise free energy by changing its configuration to change the way it samples the environment, or to change its expectations. These changes correspond to action and perception, respectively, and lead to an adaptive exchange with the environment that is characteristic of biological systems. This treatment implies that the system's state and structure encode an implicit and probabilistic model of the environment."

This area of research was summarized in terms understandable by the layperson in a 2008 article in New Scientist that offered a unifying theory of brain function. Friston makes the following claims about the explanatory power of the theory:

"This model of brain function can explain a wide range of anatomical and physiological aspects of brain systems; for example, the hierarchical deployment of cortical areas, recurrent architectures using forward and backward connections and functional asymmetries in these connections. In terms of synaptic physiology, it predicts associative plasticity and, for dynamic models, spike-timing-dependent plasticity. In terms of electrophysiology it accounts for classical and extra-classical receptive field effects and long-latency or endogenous components of evoked cortical responses. It predicts the attenuation of responses encoding prediction error with perceptual learning and explains many phenomena like repetition suppression, mismatch negativity and the P300 in electroencephalography. In psychophysical terms, it accounts for the behavioural correlates of these physiological phenomena, e.g., priming, and global precedence."

"It is fairly easy to show that both perceptual inference and learning rest on a minimisation of free energy or suppression of prediction error."

BRAIN–COMPUTER INTERFACE

A brain–computer interface (BCI), sometimes called a mind-machine interface (MMI), direct neural interface (DNI), or brain–machine interface (BMI), is a direct communication pathway between an enhanced or wired brain and an external device. BCIs are often directed at researching, mapping, assisting, augmenting, or repairing human cognitive or sensory-motor functions.

Research on BCIs began in the 1970s at the University of California, Los Angeles (UCLA) under a grant from the National Science Foundation, followed by a contract from DARPA. The papers published after this research also mark the first appearance of the expression brain–computer interface in scientific literature.

In 1980s a report was given on control of a physical object, a mobile robot, using EEG signals S. Bozinovski, M. Sestakov, L. Bozinovska: Using EEG alpha rhythm to control a mobile robot, In G. Harris, C. Walker (eds.) Proc IEEE Annual Conference on Medical and Biological Society, New Orleans, p. 1515-1516,1988S. Bozinovski: Mobile robot trajectory control: From fixed rails to direct bioelectric control, In O. Kayniak (ed.) Proc. IEEE Workshop on Intelligent Motion Control, p. 63-67, 1990.

The field of BCI research and development has since focused primarily on neuroprosthetics applications that aim at restoring damaged hearing, sight and movement. Thanks to the remarkable cortical plasticity of the brain, signals from implanted prostheses can, after adaptation, be handled by the brain like natural sensor or effector channels. Following years of animal experimentation, the first neuroprosthetic devices implanted in humans appeared in the mid-1990s.

The history of brain–computer interfaces (BCIs) starts with Hans Berger's discovery of the electrical activity of the human brain and the development of electroencephalography (EEG). In 1924 Berger was the first to record human brain activity by means of EEG. Berger was able to identify oscillatory activity, such as Berger's wave or the alpha wave (8–13 Hz), by analyzing EEG traces.

Berger's first recording device was very rudimentary. He inserted silver wires under the scalps of his patients. These were later replaced by silver foils attached to the patients' head by rubber bandages. Berger connected these sensors to a Lippmann capillary electrometer, with disappointing results. However, more sophisticated measuring devices, such as the Siemens double-coil recording galvanometer, which displayed electric voltages as small as one ten thousandth of a volt, led to success.

Berger analyzed the interrelation of alternations in his EEG wave diagrams with brain diseases. EEGs permitted completely new possibilities for the research of human brain activities.

Jacques Vidal coined the term "BCI" and produced the first peer-reviewed publications on this topic. Vidal is widely recognized as the inventor of BCIs in the BCI community, as reflected in numerous peer-reviewed articles reviewing and discussing the field (e.g.,).

After his early contributions, Vidal was not active in BCI research, nor BCI events such as conferences, for many years. In 2011, however, he gave a lecture in Graz, Austria, supported by the Future BNCI project, presenting the first BCI, which earned a standing ovation. Vidal was joined by his wife, Laryce Vidal, who previously worked with him at UCLA on his first BCI project. Prof. Vidal will also present a lecture on his early BCI work at the Sixth Annual BCI Meeting, scheduled for May–June 2016 at Asilomar, California.

Versus Neuroprosthetics

Neuroprosthetics is an area of neuroscience concerned with neural prostheses, that is, using artificial devices to replace the function of impaired nervous systems and brain related problems, or of sensory organs. The most widely used neuroprosthetic device is the cochlear implant which, as of December 2010, had been implanted in approximately 220,000 people worldwide. There are also several neuroprosthetic devices that aim to restore vision, including retinal implants.

The difference between BCIs and neuroprosthetics is mostly in how the terms are used: neuroprosthetics typically connect the nervous system to a device, whereas BCIs usually connect the brain (or nervous system) with a computer system. Practical neuroprosthetics can be linked to any part of the nervous system—for example, peripheral nerves—while the term "BCI" usually designates a narrower class of systems which interface with the central nervous system.

The terms are sometimes, however, used interchangeably. Neuroprosthetics and BCIs seek to achieve the same aims, such as restoring sight, hearing, movement, ability to communicate, and even cognitive function. Both use similar experimental methods and surgical techniques.

Animal BCI Research

Several laboratories have managed to record signals from monkey and rat cerebral cortices to operate BCIs to produce movement. Monkeys have navigated computer cursors on screen and commanded robotic arms to perform simple tasks simply by thinking about the task and seeing the visual feedback, but without any motor output. In May 2008 photographs that showed a monkey at the University of Pittsburgh Medical Center operating a robotic arm by thinking were published in a number of well-known science journals and magazines. Other research on cats has decoded their neural visual signals.

Early Work

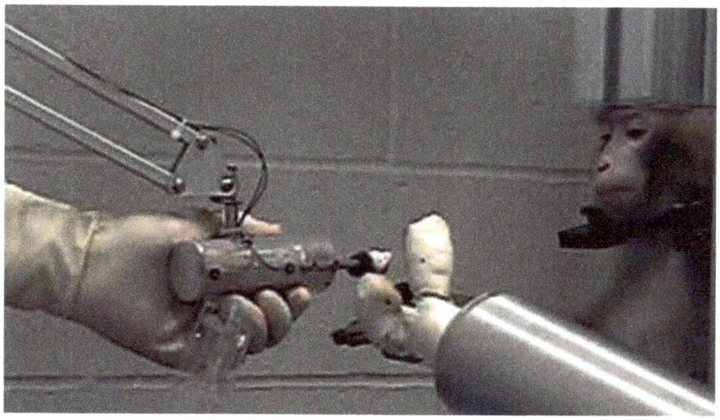

Monkey operating a robotic arm with brain–computer interfacing (Schwartz lab, University of Pittsburgh)

In 1969 the operant conditioning studies of Fetz and colleagues, at the Regional Primate Research Center and Department of Physiology and Biophysics, University of Washington School of Medicine in Seattle, showed for the first time that monkeys could learn to control the deflection of a biofeedback meter arm with neural activity. Similar work in the 1970s established that monkeys could quickly learn to voluntarily control the firing rates of individual and multiple neurons in the primary motor cortex if they were rewarded for generating appropriate patterns of neural activity.

Studies that developed algorithms to reconstruct movements from motor cortex neurons, which control movement, date back to the 1970s. In the 1980s, Apostolos Georgopoulos at Johns Hopkins University found a mathematical relationship between the electrical responses of single motor cortex neurons in rhesus macaque monkeys and the direction in which they moved their arms (based on a cosine function). He also found that dispersed groups of neurons, in different areas of the monkey's brains, collectively controlled motor commands, but was able to record the firings of neurons in only one area at a time, because of the technical limitations imposed by his equipment.

There has been rapid development in BCIs since the mid-1990s. Several groups have been able to capture complex brain motor cortex signals by recording from neural ensembles (groups of neurons) and using these to control external devices.

Prominent Research Successes

Kennedy and Yang Dan

Phillip Kennedy (who later founded Neural Signals in 1987) and colleagues built the first intracortical brain–computer interface by implanting neurotrophic-cone electrodes into monkeys.

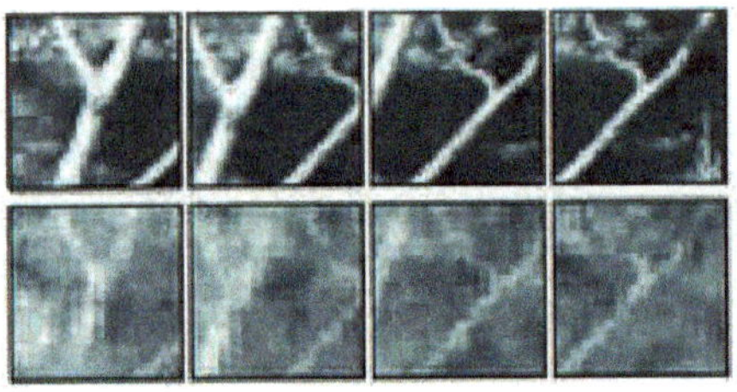

Yang Dan and colleagues' recordings of cat vision using a BCI implanted in the lateral geniculate nucleus (top row: original image; bottom row: recording)

In 1999, researchers led by Yang Dan at the University of California, Berkeley decoded neuronal firings to reproduce images seen by cats. The team used an array of electrodes embedded in the thalamus (which integrates all of the brain's sensory input) of sharp-eyed cats. Researchers targeted 177 brain cells in the thalamus lateral geniculate nucleus area, which decodes signals from the retina. The cats were shown eight short movies, and their neuron firings were recorded. Using mathematical filters, the researchers decoded the signals to generate movies of what the cats saw and were able to reconstruct recognizable scenes and moving objects. Similar results in humans have since been achieved by researchers in Japan (see below).

Nicolelis

Miguel Nicolelis, a professor at Duke University, in Durham, North Carolina, has been a prominent proponent of using multiple electrodes spread over a greater area of the brain to obtain neuronal signals to drive a BCI.

After conducting initial studies in rats during the 1990s, Nicolelis and his colleagues developed BCIs that decoded brain activity in owl monkeys and used the devices to reproduce monkey movements in robotic arms. Monkeys have advanced reaching and grasping abilities and good hand manipulation skills, making them ideal test subjects for this kind of work.

By 2000 the group succeeded in building a BCI that reproduced owl monkey movements while the monkey operated a joystick or reached for food. The BCI operated in real time and could also control a separate robot remotely over Internet protocol. But the monkeys could not see the arm moving and did not receive any feedback, a so-called open-loop BCI.

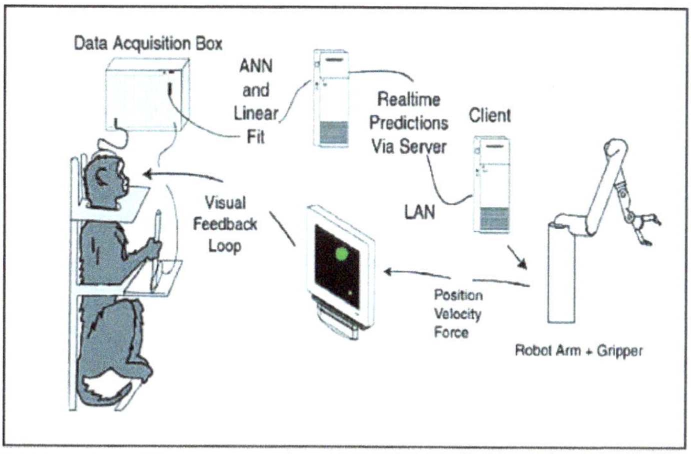

Diagram of the BCI developed by Miguel Nicolelis and colleagues for use on rhesus monkeys

Later experiments by Nicolelis using rhesus monkeys succeeded in closing the feedback loop and reproduced monkey reaching and grasping movements in a robot arm. With their deeply cleft and furrowed brains, rhesus monkeys are considered to be better models for human neurophysiology than owl monkeys. The monkeys were trained to reach and grasp objects on a computer screen by manipulating a joystick while corresponding movements by a robot arm were hidden. The monkeys were later shown the robot directly and learned to control it by viewing its movements. The BCI used velocity predictions to control reaching movements and simultaneously predicted handgripping force. In 2011 O'Doherty and colleagues showed a BCI with sensory feedback with rhesus monkeys. The monkey was brain controlling the position of an avatar arm while receiving sensory feedback through direct intracortical stimulation (ICMS) in the arm representation area of the sensory cortex.

Donoghue, Schwartz and Andersen

Other laboratories which have developed BCIs and algorithms that decode neuron signals include those run by John Donoghue at Brown University, Andrew Schwartz at the University of Pittsburgh and Richard Andersen at Caltech. These researchers have been able to produce working BCIs, even using recorded signals from far fewer neurons than did Nicolelis (15–30 neurons versus 50–200 neurons).

Donoghue's group reported training rhesus monkeys to use a BCI to track visual targets on a computer screen (closed-loop BCI) with or without assistance of a joystick. Schwartz's group created a BCI for three-dimensional tracking in virtual reality and also reproduced BCI control in a robotic arm. The same group also created headlines when they demonstrated that a monkey could feed itself pieces of fruit and marshmallows using a robotic arm controlled by the animal's own brain signals.

Andersen's group used recordings of premovement activity from the posterior parietal cortex in their BCI, including signals created when experimental animals anticipated receiving a reward.

Other Research

In addition to predicting kinematic and kinetic parameters of limb movements, BCIs that predict electromyographic or electrical activity of the muscles of primates are being developed.Such BCIs could be used to restore mobility in paralyzed limbs by electrically stimulating muscles.

Miguel Nicolelis and colleagues demonstrated that the activity of large neural ensembles can predict arm position. This work made possible creation of BCIs that read arm movement intentions and translate them into movements of artificial actuators. Carmena and colleagues programmed the neural coding in a BCI that allowed a monkey to control reaching and grasping movements by a robotic arm. Lebedev and colleagues argued that brain networks reorganize to create a new representation of the robotic appendage in addition to the representation of the animal's own limbs.

The biggest impediment to BCI technology at present is the lack of a sensor modality that provides safe, accurate and robust access to brain signals. It is conceivable or even likely, however, that such a sensor will be developed within the next twenty years. The use of such a sensor should greatly expand the range of communication functions that can be provided using a BCI.

Development and implementation of a BCI system is complex and time consuming. In response to this problem, Gerwin Schalk has been developing a general-purpose system for BCI research, called BCI2000. BCI2000 has been in development since 2000 in a project led by the Brain–Computer Interface R&D Program at the Wadsworth Center of the New York State Department of Health in Albany, New York, United States.

A new 'wireless' approach uses light-gated ion channels such as Channelrhodopsin to control the activity of genetically defined subsets of neurons in vivo. In the context of a simple learning task, illumination of transfected cells in the somatosensory cortex influenced the decision making process of freely moving mice.

The use of BMIs has also led to a deeper understanding of neural networks and the central nervous system. Research has shown that despite the inclination of neuroscientists to believe that neurons have the most effect when working together, single neurons can be conditioned through the use of BMIs to fire at a pattern that allows primates to control motor outputs. The use of BMIs has led to development of the single neuron insufficiency principle which states that even with a well tuned firing rate single neurons can only carry a narrow amount of information and therefore the highest level of accuracy is achieved by recording firings of the collective ensemble. Other principles discovered with the use of BMIs include the neuronal multitasking principle, the neuronal mass principle, the neural degeneracy principle, and the plasticity principle.

BCIs are also proposed to be applied by users without disabilities. A user-centered categorization of BCI approaches by Thorsten O. Zander and Christian Kothe introduces the term passive BCI. Next to active and reactive BCI that are used for directed control, passive BCIs allow for assessing and interpreting changes in the user state during Human-Computer Interaction (HCI). In a secondary, implicit control loop the computer system adapts to its user improving its usability in general.

The BCI Award

The Annual BCI Research Award, endowed with 3,000 USD, is awarded in recognition of outstanding and innovative research in the field of Brain-Computer Interfaces. Each year, a renowned research laboratory is asked to judge the submitted projects and to award the prize. The jury consists of world-leading BCI experts recruited by the awarding laboratory. Following list consists the winners of the BCI Award:

- 2010: Cuntai Guan, Kai Keng Ang, Karen Sui Geok Chua and Beng Ti Ang, (A*STAR, Singapore)

Motor imagery-based Brain-Computer Interface robotic rehabilitation for stroke.

- 2011: Moritz Grosse-Wentrup and Bernhard Schölkopf, (Max Planck Institute for Intelligent Systems, Germany)

What are the neuro-physiological causes of performance variations in brain-computer interfacing?

- 2012: Surjo R. Soekadar and Niels Birbaumer, (Applied Neurotechnology Lab, University Hospital Tübingen and Institute of Medical Psychology and Behavioral Neurobiology, Eberhard Karls University, Tübingen, Germany)

Improving Efficacy of Ipsilesional Brain-Computer Interface Training in Neurorehabilitation of Chronic Stroke

- 2013: M. C. Dadarlat[a,b], J. E. O'Doherty[a], P. N. Sabes[a,b] ([a]Department of Physiology, Center for Integrative Neuroscience, San Francisco, CA, US, [b]UC Berkeley-UCSF Bioengineering Graduate Program, University of California, San Francisco, CA, US)

A learning-based approach to artificial sensory feedback: intracortical microstimulation replaces and augments vision

- 2014: Katsuhiko Hamada, Hiromu Mori, Hiroyuki Shinoda, Tomasz M. Rutkowski, (The University of Tokyo, JP, Life Science Center of TARA, University of Tsukuba, JP, RIKEN Brain Science Institute, JP)

Airborne Ultrasonic Tactile Display BCI

- 2015: Guy Hotson, David P McMullen, Matthew S. Fifer, Matthew S. Johannes, Kapil D. Katyal, Matthew P. Para, Robert Armiger, William S. Anderson, Nitish V. Thakor, Brock A. Wester, Nathan E. Crone (Johns Hopkins University, USA)

Individual Finger Control of the Modular Prosthetic Limb using High-Density Electrocorticography in a Human Subject

- 2016: Gaurav Sharma, Nick Annetta, Dave Friedenberg, Marcie Bockbrader, Ammar Shaikhouni, W. Mysiw, Chad Bouton, Ali Rezai (Battelle Memorial Institute, The Ohio State University, USA)

An Implanted BCI for Real-Time Cortical Control of Functional Wrist and Finger Movements in a Human with Quadriplegia

Human BCI Research

Invasive BCIs

Vision

Jens Naumann, a man with acquired blindness, being interviewed about his vision BCI on CBS's The Early Show

Invasive BCI research has targeted repairing damaged sight and providing new functionality for people with paralysis. Invasive BCIs are implanted directly into the grey matter of the brain during neurosurgery. Because they lie in the grey matter, invasive devices produce the highest quality signals of BCI devices but are prone to scar-tissue build-up, causing the signal to become weaker, or even non-existent, as the body reacts to a foreign object in the brain.

In *vision science*, direct brain implants have been used to treat non-congenital (acquired) blindness. One of the first scientists to produce a working brain interface to restore sight was private researcher William Dobelle.

Dobelle's first prototype was implanted into "Jerry", a man blinded in adulthood, in 1978. A single-array BCI containing 68 electrodes was implanted onto Jerry's visual cortex and succeeded in producing phosphenes, the sensation of seeing light. The system included cameras mounted on glasses to send signals to the implant. Initially, the implant allowed Jerry to see shades of grey in a limited field of vision at a low frame-rate. This also required him to be hooked up to a mainframe computer, but shrinking electronics and faster computers made his artificial eye more portable and now enable him to perform simple tasks unassisted.

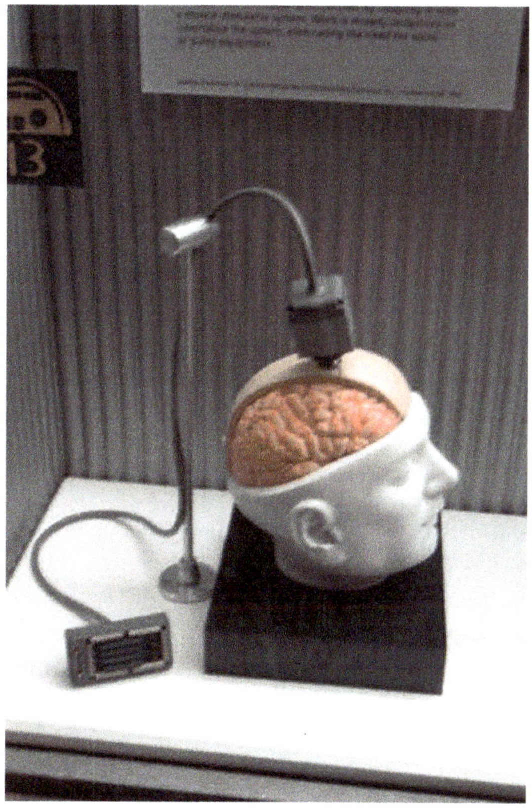

Dummy unit illustrating the design of a BrainGate interface

In 2002, Jens Naumann, also blinded in adulthood, became the first in a series of 16 paying patients to receive Dobelle's second generation implant, marking one of the earliest commercial uses of BCIs. The second generation device used a more sophisticated implant enabling better mapping of phosphenes into coherent vision. Phosphenes are spread out across the visual field in what researchers call "the starry-night effect". Immediately after his implant, Jens was able to use his imperfectly restored vision to drive an automobile slowly around the parking area of the research institute. Unfortunately, Dobelle died in 2004 before his processes and developments were documented. Subsequently, when Mr. Naumann and the other patients in the program began having problems with their vision, there was no relief and they eventually lost their "sight" again. Naumann wrote about his experience with Dobelle's work in *Search for Paradise: A Patient's Account of the Artificial Vision Experiment*and has returned to his farm in Southeast Ontario, Canada, to resume his normal activities.

Movement

BCIs focusing on *motor neuroprosthetics* aim to either restore movement in individuals with paralysis or provide devices to assist them, such as interfaces with computers or robot arms.

Researchers at Emory University in Atlanta, led by Philip Kennedy and Roy Bakay, were first to install a brain implant in a human that produced signals of high enough quality to simulate movement. Their patient, Johnny Ray (1944–2002), suffered from 'locked-in syndrome' after suffering a brain-stem stroke in 1997. Ray's implant was installed in 1998 and he lived long enough to start working with the implant, eventually learning to control a computer cursor; he died in 2002 of a brain aneurysm.

Tetraplegic Matt Nagle became the first person to control an artificial hand using a BCI in 2005 as part of the first nine-month human trial of Cyberkinetics's BrainGate chip-implant. Implanted in Nagle's right precentral gyrus (area of the motor cortex for arm movement), the 96-electrode BrainGate implant allowed Nagle to control a robotic arm by thinking about moving his hand as well as a computer cursor, lights and TV. One year later, professor Jonathan Wolpaw received the prize of the Altran Foundation for Innovation to develop a Brain Computer Interface with electrodes located on the surface of the skull, instead of directly in the brain.

More recently, research teams led by the Braingate group at Brown University and a group led by University of Pittsburgh Medical Center, both in collaborations with the United States Department of Veterans Affairs, have demonstrated further success in direct control of robotic prosthetic limbs with many degrees of freedom using direct connections to arrays of neurons in the motor cortex of patients with tetraplegia.

Partially Invasive BCIs

Partially invasive BCI devices are implanted inside the skull but rest outside the brain rather than within the grey matter. They produce better resolution signals than non-invasive BCIs where the bone tissue of the cranium deflects and deforms signals and have a lower risk of forming scar-tissue in the brain than fully invasive BCIs. There has been preclinical demonstration of intracortical BCIs from the stroke perilesional cortex.

Electrocorticography (ECoG) measures the electrical activity of the brain taken from beneath the skull in a similar way to non-invasive electroencephalography (see below), but the electrodes are embedded in a thin plastic pad that is placed above the cortex, beneath the dura mater. ECoG technologies were first trialled in humans in 2004 by Eric Leuthardt and Daniel Moran from Washington University in St Louis. In a later trial, the researchers enabled a teenage boy to play Space Invaders using his ECoG implant. This research indicates that control is rapid, requires minimal training, and may be an ideal tradeoff with regards to signal fidelity and level of invasiveness.

(Note: these electrodes had not been implanted in the patient with the intention of developing a BCI. The patient had been suffering from severe epilepsy and the electrodes were temporarily implanted to help his physicians localize seizure foci; the BCI researchers simply took advantage of this.)

Signals can be either subdural or epidural, but are not taken from within the brain parenchyma itself. It has not been studied extensively until recently due to the limited access of subjects. Currently, the only manner to acquire the signal for study is through the use of patients requiring invasive monitoring for localization and resection of an epileptogenic focus.

ECoG is a very promising intermediate BCI modality because it has higher spatial resolution, better signal-to-noise ratio, wider frequency range, and less training requirements than scalp-recorded EEG, and at the same time has lower technical difficulty, lower clinical risk, and probably superior long-term stability than intracortical single-neuron recording. This feature profile and recent evidence of the high level of control with minimal training requirements shows potential for real world application for people with motor disabilities.

Light reactive imaging BCI devices are still in the realm of theory. These would involve implanting a laser inside the skull. The laser would be trained on a single neuron and the neuron›s reflectance measured by a separate sensor. When the neuron fires, the laser light pattern and wavelengths it reflects would change slightly. This would allow researchers to monitor single neurons but require less contact with tissue and reduce the risk of scar-tissue build-up.

In 2014, a BCI study using near-infrared spectroscopy for "locked-in" patients with amyotrophic lateral sclerosis (ALS) was able to restore some basic ability of the patients to communicate with other people.

Non-Invasive BCIs

There have also been experiments in humans using non-invasive neuroimaging technologies as interfaces. The substantial majority of published BCI work involves noninvasive EEG-based BCIs. Noninvasive EEG-based technologies and interfaces have been used for a much broader variety of applications. Although EEG-based interfaces are easy to wear and do not require surgery, they have relatively poor spatial resolution and cannot effectively use higher-frequency signals because the skull dampens signals, dispersing and blurring the electromagnetic waves created by the neurons. EEG-based interfaces also require some time and effort prior to each usage session, whereas non-EEG-based ones, as well as invasive ones require no prior-usage training. Overall, the best BCI for each user depends on numerous factors.

Non-EEG-Based Human-Computer Interface

Pupil-Size Oscillation

In a recent 2016 article, an entirely new communication device and non-EEG-based human-computer interface was developed, requiring no visual fixation or ability to move eyes at all, that is based on covert interest in (i.e. without fixing eyes on) chosen letter on a virtual keyboard with letters each having its own (background) circle that is micro-oscillating in brightness in different time transitions, where the letter selection is based on best fit between, on one hand, unintentional pupil-size oscillation pattern, and, on the other hand, the circle-in-background's brightness oscillation pattern. Accuracy is additionally improved by user's mental rehearsing the words 'bright' and 'dark' in synchrony with the brightness transitions of the circle/letter.

Electroencephalography (EEG)-Based Brain-Computer Interfaces

Overview

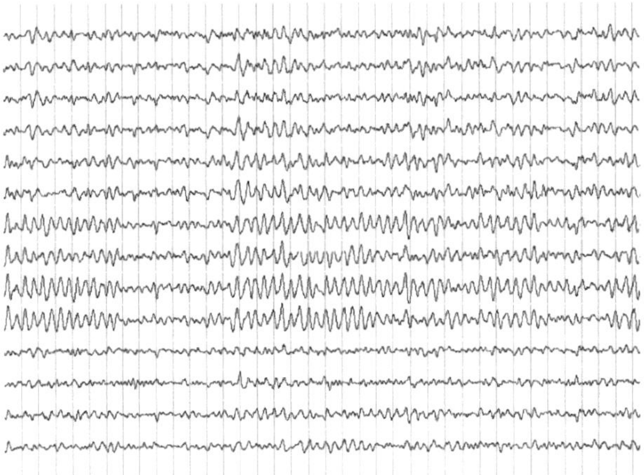

Recordings of brainwaves produced by an electroencephalogram

Electroencephalography (EEG) is the most studied non-invasive interface, mainly due to its fine temporal resolution, ease of use, portability and low set-up cost. The technology is somewhat susceptible to noise however.

In the early days of BCI research, another substantial barrier to using EEG as a brain–computer interface was the extensive training required before users can work the technology. For example, in experiments beginning in the mid-1990s, Niels Birbaumer at the University of Tübingen in Germany trained severely paralysed people to self-regulate the *slow cortical potentials* in their EEG to such an extent that these signals could be used as a binary signal to control a computer cursor. (Birbaumer had earlier trained epileptics to prevent impending fits by controlling this low voltage wave.) The experiment saw ten patients trained to move a computer cursor by controlling their brainwaves. The process was slow, requiring more than an hour for patients to write 100 characters with the cursor, while training often took many months. However, the slow cortical potential approach to BCIs has not been used in several years, since other approaches require little or no training, are faster and more accurate, and work for a greater proportion of users.

Another research parameter is the type of oscillatory activity that is measured. Birbaumer›s later research with Jonathan Wolpaw at New York State University has focused on developing technology that would allow users to choose the brain signals they found easiest to operate a BCI, including *mu* and *beta* rhythms.

A further parameter is the method of feedback used and this is shown in studies of P300 signals. Patterns of P300 waves are generated involuntarily (stimulus-feedback) when people see something they recognize and may allow BCIs to decode categories of thoughts without training patients first. By contrast, the biofeedback methods described above require learning to control brainwaves so the resulting brain activity can be detected.

While an EEG based brain-computer interface has been pursued extensively by a number of research labs, recent advancements made by Bin He and his team at the University of Minnesota suggest the potential of an EEG based brain-computer interface to accomplish tasks close to invasive brain-computer interface. Using advanced functional neuroimaging including BOLD functional MRI and EEG source imaging, Bin He and co-workers identified the co-variation and co-localization of electrophysiological and hemodynamic signals induced by motor imagination. Refined by a neuroimaging approach and by a training protocol, Bin He and co-workers demonstrated the ability of a non-invasive EEG based brain-computer interface to control the flight of a virtual helicopter in 3-dimensional space, based upon motor imagination. In June 2013 it was announced that Bin He had developed the technique to enable a remote-control helicopter to be guided through an obstacle course.

In addition to a brain-computer interface based on brain waves, as recorded from scalp EEG electrodes, Bin He and co-workers explored a virtual EEG signal-based brain-computer interface by first solving the EEG inverse problem and then used the resulting virtual EEG for brain-computer interface tasks. Well-controlled studies suggested the merits of such a source analysis based brain-computer interface.

A 2014 study found that severely motor-impaired patients could communicate faster and more reliably with non-invasive EEG BCI, than with any muscle-based communication channel.

Dry active Electrode Arrays

In the early 1990s Babak Taheri, at University of California, Davis demonstrated the first single and also multichannel dry active electrode arrays using micro-machining. The single channel dry EEG electrode construction and results were published in 1994. The arrayed electrode was also demonstrated to perform well compared to silver/silver chloride electrodes. The device consisted of four sites of sensors with integrated electronics to reduce noise by impedance matching. The advantages of such electrodes are: (1) no electrolyte used, (2) no skin preparation, (3) significantly reduced sensor size, and (4) compatibility with EEG monitoring systems. The active electrode array is an integrated system made of an array of capacitive sensors with local integrated circuitry housed in a package with batteries to power the circuitry. This level of integration was required to achieve the functional performance obtained by the electrode.

The electrode was tested on an electrical test bench and on human subjects in four modalities of EEG activity, namely: (1) spontaneous EEG, (2) sensory event-related potentials, (3) brain stem potentials, and (4) cognitive event-related potentials. The performance of the dry electrode compared favorably with that of the standard wet electrodes in terms of skin preparation, no gel requirements (dry), and higher signal-to-noise ratio.

In 1999 researchers at Case Western Reserve University, in Cleveland, Ohio, led by Hunter Peckham, used 64-electrode EEG skullcap to return limited hand movements to quadriplegic Jim Jatich. As Jatich concentrated on simple but opposite concepts like up and down, his beta-rhythm EEG output was analysed using software to identify patterns in the noise. A basic pattern was identified and used to control a switch: Above average activity was set to on, below average off. As well as enabling Jatich to control a computer cursor the signals were also used to drive the nerve controllers embedded in his hands, restoring some movement.

Prosthesis and Environment Control

Non-invasive BCIs have also been applied to enable brain-control of prosthetic upper and lower extremity devices in people with paralysis. For example, Gert Pfurtscheller of Graz University of Technology and colleagues demonstrated a BCI-controlled functional electrical stimulation system to restore upper extremity movements in a person with tetraplegia due to spinal cord injury. Between 2012 and 2013, researchers at the University of California, Irvine demonstrated for the first time that it is possible to use BCI technology to restore brain-controlled walking after spinal cord injury. In their spinal cord injury research study, a person with paraplegia was able to operate a BCI-robotic gait orthosis to regain basic brain-controlled ambulation. In 2009 Alex Blainey, an independent researcher based in the UK, successfully used the Emotiv EPOC to control a 5 axis robot arm. He then went on to make several demonstration mind controlled wheelchairs and home automation that could be operated by people with limited or no motor control such as those with paraplegia and cerebral palsy.

Other Research

Electronic neural networks have been deployed which shift the learning phase from the user to the computer. Experiments by scientists at the Fraunhofer Society in 2004 using neural networks led to noticeable improvements within 30 minutes of training.

Experiments by Eduardo Miranda, at the University of Plymouth in the UK, has aimed to use EEG recordings of mental activity associated with music to allow the disabled to express themselves musically through an encephalophone. Ramaswamy Palaniappan has pioneered the development of BCI for use in biometrics to identify/authenticate a person. The method has also been suggested for use as PIN generation device (for example in ATM and internet banking transactions. The group which is now at University of Wolverhampton has previously developed analogue cursor control using thoughts. Researchers at the University of Twente in the Netherlands have been conducting research on using BCIs for non-disabled individuals, proposing that BCIs could improve error handling, task performance, and user experience and that they could broaden the user spectrum. They particularly focused on BCI games, suggesting that BCI games could provide challenge, fantasy and sociality to game players and could, thus, improve player experience.

The first BCI session with 100% accuracy (based on 80 right-hand and 80 left-hand movement imaginations) was recorded in 1998 by Christoph Guger. The BCI system used 27 electrodes overlaying the sensorimotor cortex, weighted the electrodes with Common Spatial Patterns, calculated the running variance and used a linear discriminant analysis.

Research is ongoing into military use of BCIs and since the 1970s DARPA has been funding research on this topic. The current focus of research is user-to-user communication through analysis of neural signals. The project "Silent Talk" aims to detect and analyze the word-specific neural signals, using EEG, which occur before speech is vocalized, and to see if the patterns are generalizable.

DIY and Open Source BCI

In 2001, The OpenEEG Project was initiated by a group of DIY neuroscientists and engineers. The ModularEEG was the primary device created the OpenEEG community; it was a 6-channel signal capture board that cost between $200 and $400 to make at home. The OpenEEG Project marked a significant moment in the emergence of DIY brain-computer interfacing.

In 2010, the Frontier Nerds of NYU's ITP program published a thorough tutorial titled How To Hack Toy EEGs. The tutorial, which stirred the minds of many budding DIY BCI enthusiasts, demonstrated how to create a single channel at-home EEG with an Arduino and a Mattel Mindflex at a very reasonable price. This tutorial amplified the DIY BCI movement.

In 2013, OpenBCI emerged from a DARPA solicitation and subsequent Kickstarter campaign. They created a high-quality, open-source 8-channel EEG acquisition board, known as the 32bit Board, that retailed for under $500. Two years later they created the first 3D-printed EEG Headset, known as the Ultracortex, as well as, a 4-channel EEG acquisition board, known as the Ganglion Board, that retailed for under $100.

In 2015, NeuroTechX was created with the mission of building an international network for neurotechnology. They bring hackers, researchers and enthusiasts all together in many different cities around the world. According to their rapid growth, the DIY neurotech / BCI community was already waiting for such initiative to see light.

MEG and MRI

Magnetoencephalography (MEG) and functional magnetic resonance imaging (fMRI) have both been used successfully as non-invasive BCIs. In a widely reported experiment, fMRI allowed two users being scanned to play Pong in real-time by altering their haemodynamic response or brain blood flow through biofeedback techniques.

fMRI measurements of haemodynamic responses in real time have also been used to control robot arms with a seven-second delay between thought and movement.

In 2008 research developed in the Advanced Telecommunications Research (ATR) Computational Neuroscience Laboratories in Kyoto, Japan, allowed the scientists to reconstruct images directly from the brain and display them on a computer in black and white at a resolution of 10x10 pixels. The article announcing these achievements was the cover story of the journal Neuron of 10 December 2008.

In 2011 researchers from UC Berkeley published a study reporting second-by-second reconstruction of videos watched by the study's subjects, from fMRI data. This was achieved by creating a statistical model relating visual patterns in videos shown to the subjects, to the brain activity caused by watching the videos. This model was then used to look up the 100 one-second video segments, in a database of 18 million seconds of random YouTube videos, whose visual patterns most closely matched the brain activity recorded when subjects watched a new video. These 100 one-second video extracts were then combined into a mashed-up image that resembled the video being watched.

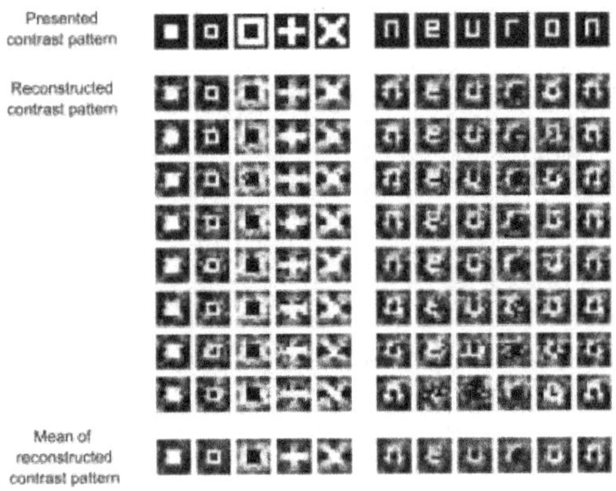

ATR Labs' reconstruction of human vision using fMRI (top row: original image; bottom row: reconstruction from mean of combined readings)

Neurogaming

Currently, there is a new field of gaming called Neurogaming, which uses non-invasive BCI in order to improve gameplay so that users can interact with a console without the use of a traditional controller. Some Neurogaming software use a player's brain waves, heart rate, expressions, pupil dilation, and even emotions to complete tasks or affect the mood of the game. For example, game developers at Emotiv have created non-invasive BCI that will determine the mood of a player and adjust music or scenery accordingly. This new form of interaction between player and software will enable a player to have a more realistic gaming experience. Because there will be less disconnect between a player and console, Neurogaming will allow individuals to utilize their "psychological state" and have their reactions transfer to games in real-time.

However, since Neurogaming is still in its first stages, not much is written about the new industry. The first NeuroGaming Conference was held in San Francisco on 1–2 May 2013.

BCI Control Strategies in Neurogaming

Motor Imagery

Motor imagery involves the imagination of the movement of various body parts resulting in sensorimotor cortex activation,which modulates sensorimotor oscillations in the EEG. This can be detected by the BCI to infer a user's intent. Motor imagery typically requires a number of sessions of training before acceptable control of the BCI is acquired. These training sessions may take a number of hours over several days before users can consistently employ the technique with acceptable levels of precision. Regardless of the duration of the training session, users are unable to master the control scheme. This results in very slow pace of the gameplay. Advance machine learning methods were recently developed to compute a subject-specific model for detecting the performance of motor imagery. The top performing algorithm from BCI Competition IV <http://www.bbci.de/competition/iv/> dataset 2 for motor imagery is the Filter Bank Common Spatial Pattern, developed by Ang et al. from A*STAR, Singapore).

Bio/Neurofeedback for Passive BCI Designs

Biofeedback is used to monitor a subject's mental relaxation. In some cases, biofeedback does not monitor electroencephalography (EEG), but instead bodily parameters such as electromyography(EMG), galvanic skin resistance (GSR), and heart rate variability (HRV).Many biofeedback systems are used to treat certain disorders such as attention deficit hyperactivity disorder (ADHD), sleep problems in children, teeth grinding, and chronic pain. EEG biofeedback systems typically monitor four different bands (theta: 4–7 Hz, alpha:8–12 Hz, SMR: 12–15 Hz, beta: 15–18 Hz) and challenge the subject to control them. Passive BCI involves using BCI to enrich human–machine interaction with implicit information on the actual user's state, for example, simulations to detect when users intend to push brakes during an emergency car stopping procedure. Game developers using passive BCIs need to acknowledge that through repetition of game levels the user's cognitive state will change or adapt. Within the first play of a level, the user will react to things differently from during the second play: for example, the user will be less surprised at an event in the game if he/she is expecting it.

Visual Evoked Potential (VEP)

A VEP is an electrical potential recorded after a subject is presented with a type of visual stimuli. There are several types of VEPs.

Steady-state visually evoked potentials (SSVEPs) use potentials generated by exciting the retina, using visual stimuli modulated at certain frequencies. SSVEP's stimuli are often formed from alternating checkerboard patterns and at times simply use flashing images . The frequency of the phase reversal of the stimulus used can be clearly distinguished in the spectrum of an EEG; this makes detection of SSVEP stimuli relatively easy . SSVEP has proved to be successful within many BCI systems . This is due to several factors, the signal elicited is measurable in as large a population as the transient VEP and blink movement and electro cardiographic artefacts do not affect the frequencies monitored. In addition, the SSVEP signal is exceptionally robust; the topographic organization of the primary visual cortex is such that a broader area obtains afferents from the central or fovial region of the visual field .SSVEP does have several problems however. As SSVEPs use flashing stimuli to infer a user's intent, the user must gaze at one of the flashing or iterating symbols in order to interact with the system. It is, therefore, likely that the symbols could become irritating and uncomfortable to use during longer play sessions, which can often last more than an hour which may not be an ideal gameplay.

Another type of VEP used with applications is the P300 potential. The P300 event-related potential is a positive peak in the EEG that occurs at roughly 300 ms after the appearance of a target stimulus (a stimulus for which the user is waiting or seeking) or oddball stimuli . The P300 amplitude decreases as the target stimuli and the ignored stimuli grow more similar.The P300 is thought to be related to a higher level attention process or an orienting response Using P300 as a control scheme has the advantage of the participant only having to attend limited training sessions. The first application to use the P300 model was the P300 matrix . Within this system, a subject would choose a letter from a grid of 6 by 6 letters and numbers. The rows and columns of the grid flashed sequentially and every time the selected "choice letter" was illuminated the user's P300 was (potentially) elicited. However, the communication process, at approximately 17 characters per minute, was quite slow. The P300 is a BCI that offers a discrete selection rather than a continuous control mechanism. The advantage of P300 use within games is that the player does not have to teach himself/herself how to use a completely new control system and so only has to undertake short training instances, to learn the gameplay mechanics and basic use of the BCI paradigm.

Synthetic Telepathy/Silent Communication

In a $6.3 million Army initiative to invent devices for telepathic communication, Gerwin Schalk, underwritten in a $2.2 million grant, found that it is possible to use ECoG signals to discriminate the vowels and consonants embedded in spoken and in imagined words. The results shed light on the distinct mechanisms associated with production of vowels and consonants, and could provide the basis for brain-based communication using imagined speech. Research into synthetic telepathy using subvocalization is taking place at the University of California, Irvine under lead scientist Mike D'Zmura. The first such communication took place in the 1960s using EEG to create Morse code using brain alpha waves. Using EEG to communicate imagined speech is less accurate than the invasive method of placing an electrode between the skull and the brain. On 27 February 2013 the group of Miguel Nicolelis at Duke University and IINN-ELS successfully connected the brains of two rats with electronic interfaces that allowed them to directly share information, in the first-ever direct brain-to-brain interface.

On 3 September 2014, scientists reported that direct communication between human brains was possible over extended distances through Internet transmission of EEG signals.

In March and in May 2014 a study conducted by Dipartimento di Psicologia Generale – Università di Padova, EVANLAB – Firenze, LiquidWeb s.r.l. company and Dipartimento di Ingegneria e Architettura – Università di Trieste, reports confirmatory results analyzing the EEG activity of two human partners spatially separated approximately 190 km apart when one member of the pair receives the stimulation and the second one is connected only mentally with the first.

Cell-Culture BCIs

Researchers have built devices to interface with neural cells and entire neural networks in cultures outside animals. As well as furthering research on animal implantable devices, experiments on cultured neural tissue have focused on building problem-solving networks, constructing basic computers and manipulating robotic devices. Research into techniques for stimulating and recording from individual neurons grown on semiconductor chips is sometimes referred to as neuroelectronics or neurochips.

The world's first Neurochip, developed by Caltech researchers Jerome Pine and Michael Maher

Development of the first working neurochip was claimed by a Caltech team led by Jerome Pine and Michael Maher in 1997. The Caltech chip had room for 16 neurons.

In 2003 a team led by Theodore Berger, at the University of Southern California, started work on a neurochip designed to function as an artificial or prosthetic hippocampus. The neurochip was designed to function in rat brains and was intended as a prototype for the eventual development of higher-brain prosthesis. The hippocampus was chosen because it is thought to be the most ordered and structured part of the brain and is the most studied area. Its function is to encode experiences for storage as long-term memories elsewhere in the brain.

In 2004 Thomas DeMarse at the University of Florida used a culture of 25,000 neurons taken from a rat's brain to fly a F-22 fighter jet aircraft simulator. After collection, the cortical neurons were cultured in a petri dish and rapidly began to reconnect themselves to form a living neural network. The cells were arranged over a grid of 60 electrodes and used to control the pitch and yaw functions of the simulator. The study's focus was on understanding how the human brain performs and learns computational tasks at a cellular level.

Ethical Considerations

Important ethical, legal and societal issues related to brain-computer interfacing are:

- conceptual issues (researchers disagree over what is and what is not a brain-computer interface),
- obtaining informed consent from people who have difficulty communicating,
- risk/benefit analysis,
- shared responsibility of BCI teams (e.g. how to ensure that responsible group decisions can be made),
- the consequences of BCI technology for the quality of life of patients and their families,

- side-effects (e.g. neurofeedback of sensorimotor rhythm training is reported to affect sleep quality),
- personal responsibility and its possible constraints (e.g. who is responsible for erroneous actions with a neuroprosthesis),
- issues concerning personality and personhood and its possible alteration,
- therapeutic applications and their possible exceedance,
- questions of research ethics that arise when progressing from animal experimentation to application in human subjects,
- mind-reading and privacy,
- mind-control,
- use of the technology in advanced interrogation techniques by governmental authorities,
- selective enhancement and social stratification.
- communication to the media.

Clausen stated in 2009 that "BCIs pose ethical challenges, but these are conceptually similar to those that bioethicists have addressed for other realms of therapy". Moreover, he suggests that bioethics is well-prepared to deal with the issues that arise with BCI technologies. Haselager and colleagues pointed out that expectations of BCI efficacy and value play a great role in ethical analysis and the way BCI scientists should approach media. Furthermore, standard protocols can be implemented to ensure ethically sound informed-consent procedures with locked-in patients. Researchers are well aware that sound ethical guidelines, appropriately moderated enthusiasm in media coverage and education about BCI systems will be of utmost importance for the societal acceptance of this technology. Thus, recently more effort is made inside the BCI community to create consensus on ethical guidelines for BCI research, development and dissemination.

Clinical and Research-Grade BCI-Based Interfaces

Some companies have been producing high-end systems that have been widely used in established BCI labs for several years. These systems typically entail more channels than the low-cost systems below, with much higher signal quality and robustness in real-world settings. Some systems from new companies have been gaining attention for new BCI applications for new user groups, such as persons with stroke or coma.

- In 2011, Nuamps EEG from www.neuroscan.com was used to study the extent of detectable brain signals from stroke patients who performed motor imagery using BCI in a large clinical trial, and the results showed that majority of the patients (87%) could use the BCI.
- In March 2012 g.tec introduced the intendiX-SPELLER, the first commercially available BCI system for home use which can be used to control computer games and apps. It can detect different brain signals with an accuracy of 99%. has hosted several workshop tours to demonstrate the intendiX system and other hardware and software to the public, such as a workshop tour of the US West

Coast during September 2012.

- In 2012 an Italian startup company, Liquidweb s.r.l., released "Braincontrol", a first prototype of an AAC BCI-based, designed for patients in locked-in state. It was validated from 2012 and 2014 with the involvement of LIS and CLIS patients. In 2014 the company introduced the commercial version of the product, with the CE mark class I as medical device.

Low-Cost BCI-Based Interfaces

Recently a number of companies have scaled back medical grade EEG technology (and in one case, NeuroSky, rebuilt the technology from the ground up) to create inexpensive BCIs. This technology has been built into toys and gaming devices; some of these toys have been extremely commercially successful like the NeuroSky and Mattel MindFlex.

- In 2006 Sony patented a neural interface system allowing radio waves to affect signals in the neural cortex.
- In 2007 NeuroSky released the first affordable consumer based EEG along with the game NeuroBoy. This was also the first large scale EEG device to use dry sensor technology.
- In 2008 OCZ Technology developed a device for use in video games relying primarily on electromyography.
- In 2008 the Final Fantasy developer Square Enix announced that it was partnering with NeuroSky to create a game, Judecca.
- In 2009 Mattel partnered with NeuroSky to release the Mindflex, a game that used an EEG to steer a ball through an obstacle course. By far the best selling consumer based EEG to date.
- In 2009 Uncle Milton Industries partnered with NeuroSky to release the Star Wars Force Trainer, a game designed to create the illusion of possessing The Force.
- In 2009 Emotiv released the EPOC, a 14 channel EEG device that can read 4 mental states, 13 conscious states, facial expressions, and head movements. The EPOC is the first commercial BCI to use dry sensor technology, which can be dampened with a saline solution for a better connection.
- In November 2011 Time Magazine selected "necomimi" produced by Neurowear as one of the best inventions of the year. The company announced that it expected to launch a consumer version of the garment, consisting of cat-like ears controlled by a brain-wave reader produced by NeuroSky, in spring 2012.
- In February 2014 They Shall Walk (a nonprofit organization fixed on constructing exoskeletons, dubbed LIFESUITs, for paraplegics and quadriplegics) began a partnership with James W. Shakarji on the development of a wireless BCI.

Future Directions

A consortium consisting of 12 European partners has completed a roadmap to support the European Commission in their funding decisions for the new framework program Horizon

2020. The project, which was funded by the European Commission, started in November 2013 and ended in April 2015. The roadmap is now complete, and can be downloaded on the project's webpage. A 2015 publication describes some of the analyses and achievements of this project, as well as the emerging Brain-Computer Interface Society. For example, this article reviewed work within this project that further defined BCIs and applications, explored recent trends, discussed ethical issues, and evaluated different directions for new BCIs. As the article notes, their new roadmap generally extends and supports the recommendations from the Future BNCI project, which conveys some enthusiasm for emerging BCI directions.

In addition to, other recent publications have explored the most promising future BCI directions for new groups of disabled users (e.g.,). Some prominent examples are summarized below.

Disorders of Consciousness (DOC)

Some persons have a disorder of consciousness (DOC). This state is defined to include persons with coma, as well as persons in a vegetative state (VS) or minimally conscious state (MCS). New BCI research seeks to help persons with DOC in different ways. A key initial goal is to identify patients who are able to perform basic cognitive tasks, which would of course lead to a change in their diagnosis. That is, some persons who are diagnosed with DOC may in fact be able to process information and make important life decisions (such as whether to seek therapy, where to live, and their views on end-of-life decisions regarding them). Some persons who are diagnosed with DOC die as a result of end-of-life decisions, which may be made by family members who sincerely feel this is in the patient's best interests. Given the new prospect of allowing these patients to provide their views on this decision, there would seem to be a strong ethical pressure to develop this research direction to guarantee that DOC patients are given an opportunity to decide whether they want to live.

These and other articles describe new challenges and solutions to use BCI technology to help persons with DOC. One major challenge is that these patients cannot use BCIs based on vision. Hence, new tools rely on auditory and/or vibrotactile stimuli. Patients may wear headphones and/or vibrotactile stimulators placed on the wrists, neck, leg, and/or other locations. Another challenge is that patients may fade in and out of consciousness, and can only communicate at certain times. This may indeed be a cause of mistaken diagnosis. Some patients may only be able to respond to physicians' requests during a few hours per day (which might not be predictable ahead of time) and thus may have been unresponsive during diagnosis. Therefore, new methods rely on tools that are easy to use in field settings, even without expert help, so family members and other persons without any medical or technical background can still use them. This reduces the cost, time, need for expertise, and other burdens with DOC assessment. Automated tools can ask simple questions that patients can easily answer, such as "Is your father named George?" or "Were you born in the USA?" Automated instructions inform patients that they may convey yes or no by (for example) focusing their attention on stimuli on the right vs. left wrist. This focused attention produces reliable changes in EEG patterns that can help determine that the patient is able to communicate. The results could be presented to physicians and therapists, which could lead to a revised diagnosis and therapy. In addition, these patients could then be provided with BCI-based communication tools that could help them convey basic needs, adjust bed position and HVAC (heating, ventilation, and air conditioning), and otherwise empower them to make major life decisions and communicate. This research effort was supported in part by different EU-funded projects, such as the DECODER project led by Prof. Andrea Kuebler at the University of Wuerzburg. This project contributed to the first BCI system developed for DOC assessment and communication, called mindBEAGLE. This system is designed to help non-expert users work with DOC patients, but is not intended to replace medical staff. An EU-funded project scheduled to begin in 2015 called ComAlert will conduct further research and development to improve DOC prediction, assessment, rehabilitation, and communication, called "PARC" in that project. Another project funded by the National Science Foundation is led by Profs. Dean Krusienski and Chang Nam. This project provides for improved vibrotactile systems, advanced signal analysis, and other improvements for DOC assessment and communication.

Functional Brain Mapping

Each year, about 400,000 people undergo brain mapping during neurosurgery. This procedure is often required for people with tumors or epilepsy that do not respond to medication. During this procedure, electrodes are placed on the brain to precisely identify the locations of structures and functional areas. Patients may be awake during neurosurgery and asked to perform certain tasks, such as moving fingers or repeating words. This is necessary so that surgeons can remove only the desired tissue while sparing other regions, such as critical movement or language regions. Removing too much brain tissue can cause permanent damage, while removing too little tissue can leave the underlying condition untreated and require additional neurosurgery. Thus, there is a strong need to improve both methods and systems to map the brain as effectively as possible.

In several recent publications, BCI research experts and medical doctors have collaborated to explore new ways to use BCI technology to improve neurosurgical mapping. This work focuses largely on high gamma activity, which is difficult to detect with non-invasive means. Results have led to improved methods for identifying key areas for movement, language, and other functions. A recent article addressed advances in functional brain mapping and summarizes a workshop.

BRAIN SIMULATION

There is ongoing work in the field of brain simulation, including partial and whole simulations of some animals. For example, the brains of the roundworm C. elegans, fruit fly drosophila melanogaster, mouse and rat have all been simulated to various degrees.

C. Elegans (Roundworm)

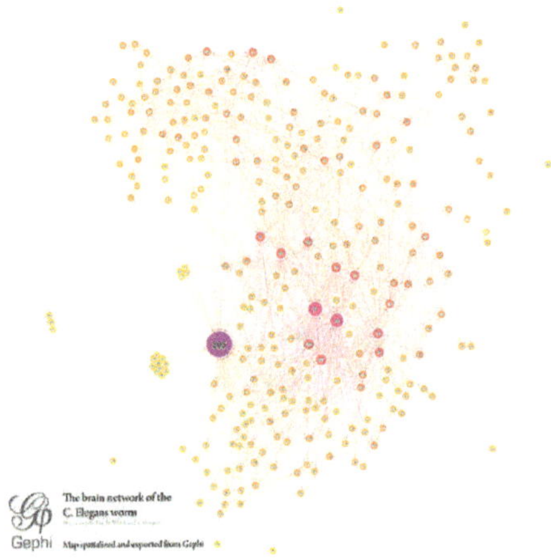

Brain map of the C. elegans roundworm 302 neurons, interconnected by 5000 synapses

The connectivity of the neural circuit for touch sensitivity of the simple C. elegans nematode (roundworm) was mapped in 1985, and partly simulated in 1993. Several software simulation models of the complete neural and muscular system, and to some extent the worm's physical environment, have been presented since 2004, and are in some cases available for downloading. However, we still lack understanding of how the neurons and the connections between them generate the surprisingly complex range of behaviors that are observed in this relatively simple organism.

Drosophila Neural System

The brain belonging to the fruit fly Drosophila is also thoroughly studied, and a simplified model simulated.

Mouse Brain Mapping and Simulation

Between 1995 and 2005, Henry Markram mapped the types of neurons and their connections in such a column.

The Blue Brain project, completed in December 2006, aimed at the simulation of a rat neocortical column, which can be considered the smallest functional unit of the neocortex (the part of the brain thought to be responsible for higher functions such as conscious thought), containing 10,000 neurons (and 10^8synapses). In November 2007, the project reported the end of the first phase, delivering a data-driven process for creating, validating, and researching the neocortical column.

An artificial neural network described as being "as big and as complex as half of a mouse brain" was run on an IBM blue gene supercomputer by a University of Nevada research team in 2007. A simulated time of one second took ten seconds of computer time. The researchers said they had seen "biologically consistent" nerve impulses flowed through the virtual cortex. However, the simulation lacked the structures seen in real mice brains, and they intend to improve the accuracy of the neuron model.

Blue Brain and the Rat

Blue Brain is a project, launched in May 2005 by IBM and the Swiss Federal Institute of Technology in Lausanne, with the aim to create a computer simulation of a mammalian cortical column, down to the molecular level. The project uses a supercomputer based on IBM's Blue Gene design to simulate the electrical behavior of neurons based upon their synaptic connectivity and complement of intrinsic membrane currents. The initial goal of the project, completed in December 2006, was the simulation of a rat neocortical column, which can be considered the smallest functional unit of the neocortex (the part of the brain thought to be responsible for higher functions such as conscious thought), containing 10,000 neurons (and 10^8 synapses). Between 1995 and 2005, Henry Markram mapped the types of neurons and their connections in such a column. In November 2007, the project reported the end of the first phase, delivering a data-driven process for creating, validating, and researching the neocortical column. The project seeks to eventually reveal aspects of human cognition and various psychiatric disorders caused by malfunctioning neurons, such as autism, and to understand how pharmacological agents affect network behavior.

COMPUTATIONAL ANATOMY

Computational anatomy is a discipline within medical imaging focusing on the study of anatomical shape and form at the visible or gross anatomical scale of morphology. It involves the development and application of computational, mathematical and data-analytical methods for modeling and simulation of biological structures.

The field is broadly defined and includes foundations in anatomy, applied mathematics and pure mathematics, machine learning, computational mechanics, computational science, medical imaging, neuroscience, physics, probability, and statistics; it also has strong connections with fluid mechanics and geometric mechanics. Additionally, it complements newer, interdisciplinary fields like bioinformatics and neuroinformatics in the sense that its interpretation uses metadata derived from the original sensor imaging modalities (of which Magnetic Resonance Imaging is one example). It focuses on the anatomical structures being imaged, rather than the medical imaging devices. It is similar in spirit to the history of Computational linguistics, a discipline that focuses on the linguistic structures rather than the sensor acting as the transmission and communication medium(s).

In computational anatomy, the diffeomorphism group is used to study different coordinate systems via coordinate transformations as generated via the Lagrangian and Eulerian velocities of flow in R^3. The flows between coordinates in Computational anatomy are constrained to be geodesic flows satisfying the principle of least action for the Kinetic energy of the flow. The kinetic energy is defined through a Sobolev smoothness norm with strictly more than two generalized, square-integrable derivatives for each component of the flow velocity, which guarantees that the flows in R^3 are diffeomorphisms. It also implies that the diffeomorphic shape momentum taken pointwise satisfying the Euler-Lagrange equation for geodesics is determined by its neighbors through spatial derivatives on the velocity field. This separates the discipline from the case of incompressible fluids for which momentum is a pointwise function of velocity. Computational anatomy intersects the study of Riemannian manifolds and nonlinear global analysis, where groups of diffeomorphisms are the central focus. Emerging high-dimensional theories of shape are central to many studies in Computational anatomy, as are questions emerging from the fledgling field of shape statistics. The metric structures in Computational anatomy are related in spirit to morphometrics, with the distinction that Computational anatomy focuses on an infinite-dimensional space of coordinate systems transformed by a diffeomorphism, hence the central use of the terminology **diffeomorphometry**, the metric space study of coordinate systems via diffeomorphisms.

Genesis

At Computational anatomy's heart is the comparison of shape by recognizing in one shape the other. This connects it to D'Arcy Wentworth Thompson's developments On Growth and Form which has led to scientific explanations of morphogenesis, the process by which patterns are formed in Biology. Albrecht Durer's Four Books on Human Proportion were arguably the earliest works on Computational anatomy. The efforts of Noam Chomsky in his pioneering of Computational Linguistics inspired the original formulation of Computational anatomy as a generative model of shape and form from exemplars acted upon via transformations.

Due to the availability of dense 3D measurements via technologies such as magnetic resonance imaging (MRI), Computational anatomy has emerged as a subfield of medical imaging and bioengineering for extracting anatomical coordinate systems at the morphome scale in 3D. The spirit of this discipline shares strong overlap with areas such as computer vision and kinematics of rigid bodies, where objects are studied by analysing the groups responsible for the movement in question. Computational anatomy departs from computer vision with its focus on rigid motions, as the infinite-dimensional diffeomorphism group is central to the analysis of Biological shapes. It is a branch of the image analysis and pattern theory school at Brown University pioneered by Ulf Grenander. In Grenander's general Metric Pattern Theory, making spaces of **patterns** into a metric space is one of the fundamental operations since being able to cluster and recognize anatomical configurations often requires a metric of close and far between shapes. The diffeomorphometry metric of Computational anatomy measures how far two diffeomorphic changes of coordinates are from each other, which in turn induces a metric on the shapes and images indexed to them. The models of metric pattern theory, in particular group action on the orbit of shapes and forms is a central tool to the formal definitions in Computational anatomy.

Computational anatomy is the study of shape and form at the morphome or gross anatomy millimeter, or morphology scale, focusing on the study of sub-manifolds of R^3 points, curves surfaces and subvolumes of human anatomy. An early modern computational neuro-anatomist was David Van Essen performing some of the early physical unfoldings of the human brain based on printing of a human cortex and cutting. Jean Talairach's publication of Tailarach coordinates is an important milestone at the morphome scale demonstrating the fundamental basis of local coordinate systems in studying neuroanatomy and therefore the clear link to charts of differential geometry. Concurrently, virtual mapping in Computational anatomy across high resolution dense image coordinates was already happening in Ruzena Bajcy's and Fred Bookstein's earliest developments based on Computed axial tomography and Magnetic resonance imagery. The earliest introduction of the use of flows of diffeomorphisms for transformation of coordinate systems in image analysis and medical imaging was by Christensen, Joshi, Miller, and Rabbitt.

The first formalization of Computational Anatomy as an orbit of exemplar templates under diffeomorphism group action was in the original lecture given by Grenander and Miller with that title in May 1997 at the 50th Anniversary of the Division of Applied Mathematics at Brown University, and subsequent publication. This was the basis for the strong departure from much of the previous work on advanced methods for spatial normalization and image registration which were historically built on notions of addition and basis expansion. The structure preserving transformations central to the modern field of Computational Anatomy, homeomorphisms and diffeomorphisms carry smooth submanifolds smoothly. They are generated via Lagrangian and Eulerian flows which satisfy a law of composition of functions forming the group property, but are not additive.

The original model of Computational anatomy was as the triple, $(\mathcal{G}, \mathcal{M}, \mathcal{P})$ the group $g \in \mathcal{G}$, the orbit of shapes and forms $m \in \mathcal{M}$, and the probability laws P which encode the variations of the objects in the orbit. The template or collection of templates are elements in the orbit $m_{temp} \in \mathcal{M}$ of shapes.

The Lagrangian and Hamiltonian formulations of the equations of motion of Computational Anatomy took off post 1997 with several pivotal meetings including the 1997 Luminy meetingorganized by the Azencott school at Ecole-Normale Cachan on the "Mathematics of Shape Recognition" and the 1998 Trimestre at Institute Henri Poincaré organized by David Mumford "Questions Mathématiques en Traitement du Signal et de l'Image" which catalyzed the Hopkins-Brown-ENS Cachan groups and subsequent developments and connections of Computational anatomy to developments in global analysis.

The developments in Computational Anatomy included the establishment of the Sobelev smoothness conditions on the diffeomorphometry metric to insure existence of solutions of variational problems in the space of diffeomorphisms, the derivation of the Euler-Lagrange equations characterizing geodesics through the group and associated conservation laws, the demonstration of the metric properties of the right invariant metric, the demonstration that the Euler-Lagrange equations have a well-posed initial value problem with unique solutions for all time, and with the first results on sectional curvatures for the diffeomorphometry metric in landmarked spaces. Following the Los Alamos meeting in 2002,Joshi's original large deformation singular *Landmark* solutions in Computational anatomy were connected to peaked *Solitons* or *Peakons* as solutions for the Camassa-Holm equation. Subsequently, connections were made between Computational anatomy's Euler-Lagrange equations for momentum densities for the right-invariant metric satisfying Sobolev smoothness to Vladimir Arnold's characterization of the Euler equation for incompressible flows as describing geodesics in the group of volume preserving diffeomorphisms. The first algorithms, generally termed LDDMM for large deformation diffeomorphic mapping for computing connections between landmarks in volumes and spherical manifolds,curves, currents and surfaces, volumes, tensors, varifolds, and time-series have followed. These contributions of Computational anatomy to the global analysis associated to the infinite dimensional manifolds of subgroups of the diffeomorphism group is far from trivial. The original idea of doing differential geometry, curvature and geodesics on infinite dimensional manifolds goes back to Bernhard Riemann›s Habilitation (Ueber die Hypothesen, welche der Geometrie zu Grunde liegen); the key modern book laying the foundations of such ideas in global analysis are from Michor.

The applications within Medical Imaging of Computational Anatomy continued to flourish after two organized meetings at the Institute for Pure and Applied Mathematics conferencesat University of California, Los Angeles. Computational anatomy has been useful in creating accurate models of the atrophy of the human brain at the morphome scale, as well as Cardiac templates, as well as in modeling biological systems. Since the late 1990s, computational anatomy has become an important part of developing emerging technologies for the field of medical imaging. Digital atlases are a fundamental part of modern Medical-school education and in neuroimaging research at the morphome scale. Atlas based methods and virtual textbooks which accommodate variations as in deformable templates are at the center of many neuro-image analysis platforms including Freesurfer, FSL,MRIStudio, SPM. Diffeomorphic registration, introduced in the 90›s, is now an important player with existing codes bases organized around Ants, Dartel, Demons,Lddmm, Stationarylddmm, Fastlddmm, Are Examples Of Actively Used computational codes for constructing correspondences between coordinate systems based on sparse features and dense images. Voxel-based morphometry(VBM) is an important technology built on many of these principles.

The Deformable Template Orbit Model of Computational Anatomy

The model of human anatomy is a deformable template, an orbit of exemplars under group action. Deformable template models have been central to Grenander's Metric Pattern theory, accounting for typicality via templates, and accounting for variability via transformation of the template. An orbit under group action as the representation of the deformable template is a classic formulation from differential geometry. The space of shapes are denoted $m \in \mathcal{M}$, with the group $(\mathcal{G}, \circ)$ with law of composition $\circ$; the action of the group on shapes is denoted g.m where the action of the group $g \cdot m \in \mathcal{M}, m \in \mathcal{M}$ is defined to satisfy $(g \circ g') \cdot m = g \cdot (g' \cdot m) \in \mathcal{M}.$

The orbit model m of computational anatomy is an abstract algebra $\mathcal{M} \doteq \{m = g \cdot m_{\text{temp}}, g \in \mathcal{G}\}$ - to be compared to linear algebra- since the groups act nonlinearly on the

The orbit model of computational anatomy is an abstract algebra - to be compared to linear algebra- since the groups act nonlinearly on the shapes. This is a generalization of the classical models of linear algebra, in which the set of finite dimensional $\mathbb{R}^n$ vectors are replaced by the finite-dimensional anatomical submanifolds (points, curves, surfaces and volumes) and images of them, and the $n \times n$ matrices of linear algebra are replaced by coordinate transformations based on linear and affine groups and the more general high-dimensional diffeomorphism groups.

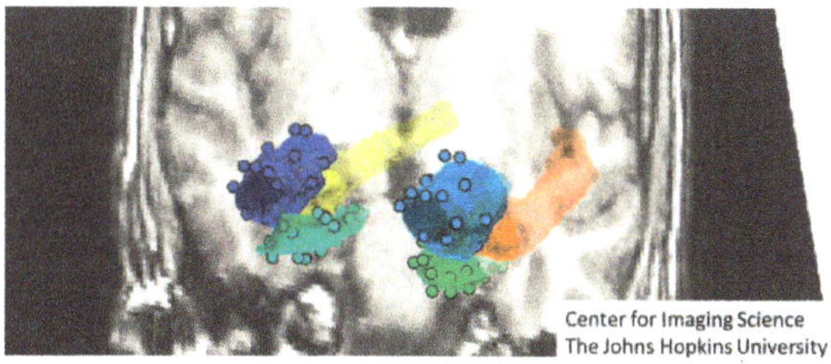

Figure depicting three medial temporal lobe structures amgydala, entorhinal cortex and hippocampus with fiducial landmarks depicted as well embedded in the MRI background.

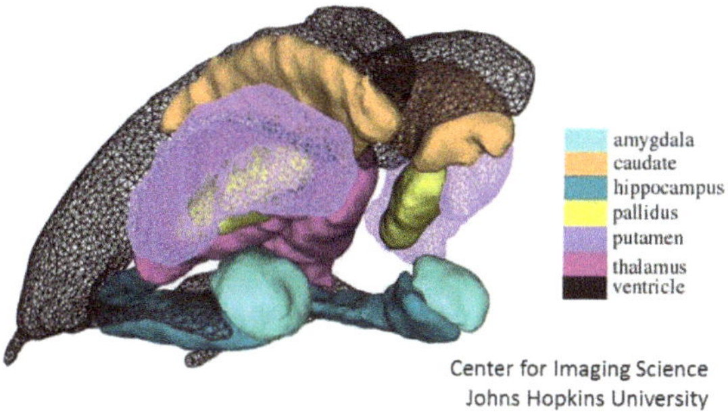

Triangulated mesh surfaces depicting subcortical structures amygdala, hippocampus, thalamus, caudate, putamen, ventricles.The shapes are denoted $m(u), u \in U \subset \mathbb{R}^1 \to \mathbb{R}^2$ represented as triangulated meshes.

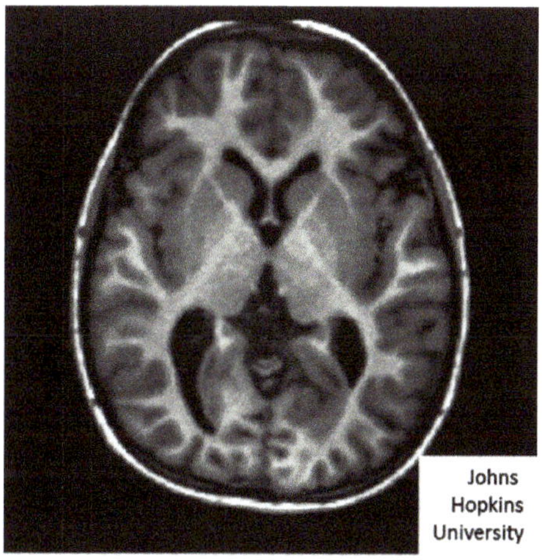

Showing an MRI section through a 3D brain representing a scalar image

$I(x), x \in \mathbb{R}^2$ based on T1-weighting.

Shapes and Forms

The central objects are shapes or forms in Computational anatomy, one set of examples being the 0,1,2,3-dimensional submanifolds of R^3 a second set of examples being images generated via medical imaging such as via magnetic resonance imaging (MRI) and functional magnetic resonance imaging.

The 0-dimensional manifolds are landmarks or fiducial points; 1-dimensional manifolds are curves such as sulcul and gyral curves in the brain; 2-dimensional manifolds correspond to boundaries of substructures in anatomy such as the subcortical structures of the midbrain or the gyral surface of the neocortex; subvolumes correspond to subregions of the human body, the heart, the thalamus, the kidney.

The landmarks $X \doteq \{x_1, \ldots, x_n\} \subset \mathbb{R}^3 \in \mathcal{M}$ are a collections of points with no other structure, delineating important fiducials within human shape and form (see associated landmarked image).

The sub-manifold shapes such as surfaces are denoted as $X \subset \mathbb{R}^3 \in \mathcal{M}$, collections of points modeled as parametrized by a local chart or immersion $m : U \subset \mathbb{R}^{1,2} \rightarrow \mathbb{R}^3$, $m(u), u \in U$

The images such as MR images or DTI images $I \in \mathcal{M}$, and are dense functions $I(x), x \in X \subset \mathbb{R}^{1,2,3}$ are scalars, vectors, and matrices (see Figure showing scalar image).

Groups and Group Actions

Groups and group actions are familiar to the Engineering community with the universal popularization and standardization of linear algebra as a basic model for analyzing signals and systems in mechanical engineering, electrical engineering and applied mathematics. In linear algebra the matrix groups (matrices with inverses) are the central structure, with group action defined by the usual definition of A as an $n \times n$ matrix, acting on $x \in \mathbb{R}^n$ as $n \times 1$ vectors; the orbit in linear-algebra is the set of n-vectors given by $y = A \cdot x \in \mathbb{R}^n$, which is a group action of the matrices through the orbit of R^n.

The central group in Computational anatomy defined on volumes in R^3 are the diffeomorphisms $\mathcal{G} \doteq Diff$ which are mappings with 3-components $\phi(\cdot) = (\phi_1(\cdot), \phi_2(\cdot), \phi_3(\cdot))$, law of composition of functions $\phi \circ \phi'(\cdot) \doteq \phi(\phi'(\cdot))$ with inverse $\phi \circ \phi^{-1}(\cdot) = \phi(\phi^{-1}(\cdot)) = id$.

Most popular are scalar images, $I(x), x \in \mathbb{R}^3$, with action on the right via the inverse.

$$\phi \cdot I(x) = I \circ \phi^{-1}(x), x \in \mathbb{R}^3.$$

For sub-manifolds $X \subset \mathbb{R}^3 \in \mathcal{M}$ parametrized by a chart or immersion $m(u), u \in U$, the diffeomorphic action the flow of the position

$$\phi \cdot m(u) \doteq \phi \circ m(u), u \in U.$$

Several group actions in computational anatomy have been defined.

Lagrangian and Eulerian Flows for Generating Diffeomorphisms

For the study of rigid body kinematics, the low-dimensional matrix Lie groups have been the central focus. The matrix groups are low-dimensional mappings, which are diffeomorphisms that provide one-to-one correspondences between coordinate systems, with a smooth inverse. The matrix group of rotations and scales can be generated via a closed form finite-dimensional matrices which are solution of simple ordinary differential equations with solutions given by the matrix exponential. For the study of deformable shape in Computational anatomy, a more general diffeomorphism group has been the group of choice, which is the infinite dimensional analogue. The highdimensional differeomorphism groups used in Computational Anatomy are generated via smooth flows $\phi_t, t \in [0,1]$ which satisfy the Lagrangian and Eulerian specification of the flow fieldssas first introduced in., satisfying the ordinary differential equation:

$$\frac{d}{dt}\phi_t = v_t \circ \phi_t,\ \phi_0 = id\,;$$

With $v \doteq (v_1, v_2, v_3)$ the vector fields on R^3 termed the Eulerian velocity of the particles at position ϕ of the flow. The vector fields are functions in a function space, modelled as a smooth Hilbert space of high-dimension, with the Jacobian of the flow $D\phi \doteq (\frac{\partial \phi_i}{\partial x_j})$ a high-dimensional field in a function space as well, rather than a low-dimensional matrix as in the matrix groups. Flows were first introduced for large deformations in image matching; $\dot{\phi}_t(x)$ is the instantaneous velocity of particle x at time t . The inverse $\phi_t^{-1}, t \in [0,1]$ required for the group is defined on the Eulerian vector-field with advective inverse flow

$$\frac{d}{dt}\phi_t^{-1} = -(D\phi_t^{-1})v_t,\ \phi_0^{-1} = id\,.$$

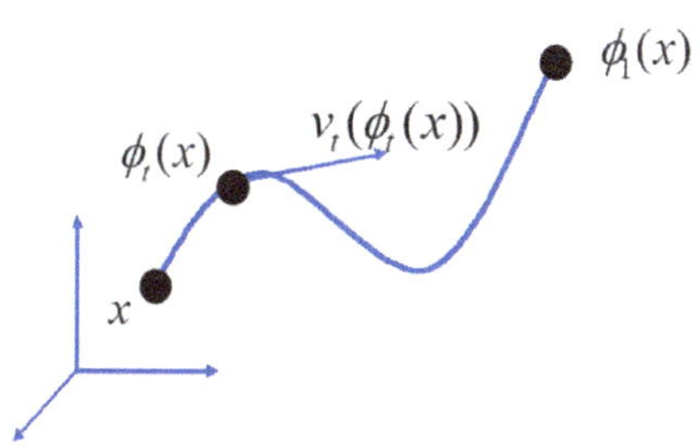

Showing the Lagrangian flow of coordinates $x \in X$ with associated vector fields $v_t, t \in [0,1]$ satisfying ordinary differential equation $\dot{\phi}_t = v_t(\phi_t), \phi_0 = id$.

The Diffeomorphism Group of Computational Anatomy

The group of diffeomorphisms is very big. To ensure smooth flows of diffeomorphisms avoiding shocklike solutions for the inverse, the vector fields must be at least 1time continuously differentiable in space. For diffeomorphisms on R^3 vector fields are modelled as elements of the Hilbert space $(V, \| \cdot \|_V)$ using the Sobolev embedding theorems so that each element $v_i \in H_0^3, i = 1, 2, 3,$ has 3-square-integrable derivatives, thusly embedding in 1-time continuously differentiable functions.

The diffeomorphism group are flows with vector fields absolutely integrable in Sobolev norm:

$$Diff_V \doteq \{\varphi = \phi_1 : \dot{\phi}_t = v_t \circ \phi_t, \phi_0 = id, \int_0^1 \|v_t\|_V dt < \infty\} .$$

Where $\|v\|_V^2 \doteq \int_X Av \cdot v dx, \; v \in V$, with the linear operator A mapping to the dual space $A : V \mapsto V^*$, with the integral calculated by integration by parts when $Av \in V^*$ is a generalized function in the dual space.

The Sobolev Smoothness Condition on Vector Fields as Modeled in a Reproducing Kernel HILBERT Space

The modelling approach used in Computational anatomy enforces a continuous differentiability condition on the vector fields by modelling the space of vector fields $(V, \| \cdot \|_V)$ as a reproducing kernel Hilbert space (RKHS), with the norm defined by a 1-1, differential operator $A : V \to V^*$. Green's inverse $K = A^{-1}$ The norm of the Hilbert space is induced by the differential operator. For $\sigma(v) \doteq Av \in V^*$ a generalized function or distribution, define the linear form as $(\sigma|w) \doteq \int_{\mathbb{R}^3} \sum_i w_i(x)\sigma_i(dx)$ This determines the norm on $(V, \| \cdot \|_V)$ according to

Since A is a differential operator, finiteness of the norm-square $(Av|v) < \infty$ includes derivatives from the differential operator implying smoothness of the vector fields. The Sobolev embedding theorem arguments were made in[72][73] demonstrating that 1-continuous derivative is required for smooth flows. For proper choice of A then $(V, \| \cdot \|_V)$ is an RKHS with the operator $K = A^{-1} : V^* \to V$ termed the Green's operator generated from the Green›s function (scalar case) for the vector field case. The Green›s kernels associated to the differential operator smooths since the kernel $k(\cdot, \cdot)$ is continuously differentiable in both variables implying

$$K\sigma(x)_i \doteq \sum_j \int_{\mathbb{R}^3} k_{ij}(x, y)\sigma_j(dy)$$

When $\sigma \doteq \mu dx$, a vector density, $(\sigma|v) \doteq \int v \cdot \mu dx$.

Diffeomorphometry: The Metric Space of Shapes and Forms

The study of metrics on groups of diffeomorphisms and the study of metrics between manifolds and surfaces has been an area of significant investigation. In Computational anatomy, the diffeomorphometry metric measures how close and far two shapes or images are from each other. Informally, the metric length is the shortest length of the flow which carries one coordinate system into the other.

Oftentimes, the familiar Euclidean metric is not directly applicable because the patterns of shapes and images don't form a vector space. In the Riemannian orbit model of Computational anatomy, diffeomorphisms acting on the forms $\phi \cdot m \in \mathcal{M}, \phi \in Diff_V, m \in \mathcal{M}$ don't act linearly. There are many ways to define metrics, and for the sets associated to shapes the Hausdorff metric is another. The method we use to induce the Riemannian metric is used to induce the metric on the orbit of shapes by defining it in terms of the metric length between diffeomorphic coordinate system transformations of the flows. Measuring the lengths of the geodesic flow between coordinates systems in the orbit of shapes is called diffeomorphometry.

The Right-invariant Metric on Diffeomorphisms

Define the distance on the group of diffeomorphisms

$$:d_{Diff_V}(\psi,\varphi) = \inf_{v_t}\left(\int_0^1 (Av_t|v_t)dt : \phi_0 = \psi, \phi_1 = \varphi, \dot{\phi}_t = v_t \circ \phi_t\right)^{1/2} ;$$

this is the right-invariant metric of diffeomorphometry, invariant to reparameterization of space since for all $\phi \in Diff_V$

$$d_{Diff_V}(\psi,\varphi) = d_{Diff_V}(\psi \circ \phi, \varphi \circ \phi).$$

The Metric on Shapes and Forms

The distance on shapes and forms, $d_{\mathcal{M}} : \mathcal{M} \times \mathcal{M} \to \mathbb{R}^+$,

$$:d_{\mathcal{M}}(m,n) = \inf_{\phi \in \mathrm{Diff}_V : \phi \cdot m = n} d_{\mathrm{Diff}_V}(id, \phi) ;$$

the images are denoted with the orbit as $I \in \mathcal{I}$ and metric , $d_{\mathcal{I}}$.

The Action Integral for Hamilton's Principle on Diffeomorphic Flows

In classical mechanics the evolution of physical systems is described by solutions to the Euler–Lagrange equations associated to the Least-action principle of Hamilton. This is a standard way, for example of obtaining Newton's laws of motion of free particles. More generally, the Euler-Lagrange equations can be derived for systems of generalized coordinates. The Euler-Lagrange equation in Computational anatomy describes the geodesic shortest path flows between coordinate systems of the diffeomorphism metric. In Computational anatomy the generalized coordinates are the flow of the diffeomorphism and its Lagrangian velocity $\phi, \dot{\phi}$, the two related via the Eulerian velocity $v \doteq \dot{\phi} \circ \phi^{-1}$ Hamilton's principle for generating the Euler-Lagrange equation requires the action integral on the Lagrangian given by

$$J(\phi) \doteq \int_0^1 L(\phi_t, \dot{\phi}_t)dt \,;$$

the Lagrangian is given by the kinetic energy:

$$L(\phi_t, \dot{\phi}_t) \doteq \frac{1}{2}\int_X A(\dot{\phi}_t \circ \phi_t^{-1}) \cdot (\dot{\phi}_t \circ \phi_t^{-1})dx = \frac{1}{2}\int_X Av_t \cdot v_t \, dx \,.$$

Diffeomorphic or Eulerian Shape Momentum

In computational anatomy, Av was first called the **Eulerian or diffeomorphic shape momentum** since when integrated against Eulerian velocity v gives energy density, and since there is a conservation of diffeomorphic shape momentum which holds. The operator A is the generalized moment of inertia or inertial operator.

The Euler–Lagrange Equation on Shape Momentum for Geodesics on the Group of Diffeomorphisms

Classical calculation of the Euler-Lagrange equation from Hamilton's principle requires the perturbation of the Lagrangian on the vector field in the kinetic energy with respect to first order perturbation of the flow. This requires adjustment by the Lie bracket of vector field, given by operator $ad_v : w \in V \mapsto V$ which involves the Jacobian given by

$$ad_v[w] \doteq [v, w] \doteq (Dv)w - (Dw)v \in V.$$

Defining the adjoint $ad_v^* : V^* \to V^*$, then the first order variation gives the Eulerian shape momentum $Av \in V^*$ satisfying the generalized equation:

$$\frac{d}{dt}Av_t + ad_{v_t}^*(Av_t) = 0 \,,\; t \in [0,1] \,;$$

meaning for all smooth $w \in V$,

$$\int_X \left(\frac{d}{dt}Av_t + ad_{v_t}^*(Av_t)\right) \cdot w dx = \int_X \frac{d}{dt}Av_t \cdot w dx + \int_X Av_t \cdot ((Dv_t)w - (Dw)v_t)dx = 0.$$

Computational anatomy is the study of the motions of submanifolds, points, curves, surfaces and volumes. Momentum associated to points, curves and surfaces are all singular, implying the momentum is concentrated on subsets of R^3 which are dimension ≤ 2 in Lebesgue measure. In such cases, the energy is still well defined $(Av_t \mid v_t)$ since although Av_t is a generalized function, the vector fields are smooth and the Eulerian momentum is understood via its action on smooth functions. The perfect illustration of this is even when it is a superposition of delta-diracs, the velocity of the coordinates in the entire volume move smoothly.The Euler-Lagrange equation (**EL-General**) on diffeomorphisms for generalized functions $Av \in V^*$ was derived in. In Riemannian Metric and Lie-Bracket Interpretation of the Euler-Lagrange Equation on Geodesics derivations are provided in terms of the adjoint operator and the Lie bracket for the group of diffeomorphisms. It has come to be called EPDiff equation for diffeomorphisms connecting to the Euler-Poincare method having been studied in the context of the inertial operator $A = identity$ for incompressible, divergence free, fluids.

Diffeomorphic Shape Momentum: a Classical Vector Function

For the momentum density case $(Av_t \mid w) = \int_X \mu_t \cdot w \, dx,$ then Euler–Lagrange equation has a classical solution:

$$\frac{d}{dt}\mu_t + (Dv_t)^T \mu_t + (D\mu_t)v_t + (\nabla \cdot v)\mu_t = 0 \,, t \in [0,1].$$

The Euler-Lagrange equation on diffeomorphisms, classically defined for momentum densities first appeared in for medical image analysis.

Riemannian Exponential (Geodesic Positioning) and Riemannian Logarithm (Geodesic Coordinates)

Global positioning systems based on systems of satellites provides a spatial navigation sytstem on the globe allowing electronic receivers to determine their location in the 3-dimensional coordinate system of longitude, latitude, and altitude to within meter scale.

In Medical imaging and Computational anatomy, positioning and coordinatizing shapes are fundamental operations; the system for positioning anatomical coordinates and shapes built on the metric and the EulerLagrange equation a geodesic positioning system as first explicated in Miller Trouve and Younes. Solving the geodesic from the initial condition v_0 is termed the Riemannian-exponential, a mapping $Exp_{id}(\cdot) : V \to Diff_V$ at identity to the group.

The Riemannian exponential satisfies $Exp_{id}(v_0) = \phi_1$ for initial condition $\dot\phi_0 = v_0$, vector field dynamics $\dot\phi_t = v_t \circ \phi_t, t \in [0,1]$, for classical equation diffeomorphic shape momentum $\int_X Av_t \cdot w \, dx$, $Av \in V$, then

$$\frac{d}{dt}Av_t + (Dv_t)^T Av_t + (DAv_t)v_t + (\nabla \cdot v)Av_t = 0 \,;$$

for generalized equation, then $Av \in V^*, w \in V$,

$$\int_X \frac{d}{dt} Av_t \cdot w dx + \int_X Av_t \cdot ((Dv_t)w - (Dw)v_t)dx = 0.$$

Computing the flow v_0 onto coordinates Riemannian logarithm. Mapping $Log_{id}(\cdot) : Diff_V \to V$ at identity from φ **to vector field** $v_0 \in V$;

$Log_{id}(\varphi) = v_0$ **initial condition of EL geodesic** $\dot{\phi}_0 = v_0, \phi_0 = id, \phi_1 = \varphi$.

Extended to the entire group they become

$\phi = Exp_\varphi(v_0 \circ \varphi) \doteq Exp_{id}(v_0) \circ \varphi$; $Log_\varphi(\phi) \doteq Log_{id}(\phi \circ \varphi^{-1}) \circ \varphi$.

These are inverses of each other for unique solutions of Logarithm; the first is called **geodesic positioning,** the latter **geodesic coordinates** (see Exponential map, Riemannian geometry for the finite dimensional version).**The geodesic metric** is a local flattening of the Riemannian coordinate system (see figure).

Hamiltonian Formulation of Computational Anatomy

In Computational anatomy the diffeomorphisms are used to push the coordinate systems, and the vector fields are used as the control within the anatomical orbit or morphological space. The model is that of a dynamical system, the flow of coordinates $t \mapsto \phi_t \in \mathrm{Diff}_V$ and the control the vector field $t \mapsto v_t \in V$ related via $\dot{\phi}_t = v_t \cdot \phi_t, \phi_0 = id.$ The Hamiltonian view reparameterizes the momentum distribution $Av \in V^*$ in terms of the conjugate momentum or canonical momentum, introduced as a Lagrange multiplier $p : \dot{\phi} \mapsto (p \mid \dot{\phi})$ constraining the Lagrangian velocity $\dot{\phi}_t = v_t \circ \phi_t$. accordingly:

$$H(\phi_t, p_t, v_t) = \int_X p_t \cdot (v_t \circ \phi_t)dx - \frac{1}{2} \int_X Av_t \cdot v_t dx.$$

This function is the extended Hamiltonian. The Pontryagin maximum principle gives the optimizing vector field which determines the geodesic flow satisfying $\dot{\phi}_t = v_t \circ \phi_t, \phi_0 = id,$ as well as the reduced Hamiltonian

$$H(\phi_t, p_t) \doteq \max_v H(\phi_t, p_t, v) .$$

The Lagrange multiplier in its action as a linear form has its own inner product of the canonical momentum acting on the velocity of the flow which is dependent on the shape, e.g. for landmarks a sum, for surfaces a surface integral, and. for volumes it is a volume integral with respect to dx on $\mathbb{R}^3$ In all cases the Greens kernels carry weights which are the canonical momentum evolving according to an ordinary differential equation which corresponds to EL but is the geodesic reparameterization in canonical momentum. The optimizing vector field is given by

$$v_t \doteq \arg max_v H(\phi_t, p_t, v)$$

with dynamics of canonical momentum reparameterizing the vector field along the geodesic

$$\begin{cases} \dot{\phi}_t = \frac{\partial H(\phi_t, p_t)}{\partial p} \\ \dot{p}_t = -\frac{\partial H(\phi_t, p_t)}{\partial \phi} \end{cases}$$

Stationarity of the Hamiltonian and Kinetic Energy Along Euler–Lagrange

Whereas the vector fields are extended across the entire background space of R^3, the geodesic flows associated to the submanifolds has Eulerian shape momentum which evolves as a generalized function $Av_t \in V^*$ concentrated to the submanifolds. For landmarks the geodesics have Eulerian shape momentum which are a superposition of delta distributions travelling with the finite numbers of particles; the diffeomorphic flow of coordinates have velocities in the range of weighted Green's Kernels. For surfaces, the momentum is a surface integral of delta distributions travelling with the surface.

The geodesics connecting coordinate systems satisfying **EL-General** have stationarity of the Lagrangian. The Hamiltonian is given by the extremum along the path $t \in [0,1]$, $H(\phi, p) = \max_v H(\phi, p, v)$, equalling the Lagrangian-Kinetic-Energy and is stationary along EL-General. Defining the geodesic velocity at the identity $v_0 = \arg\max_v H(\phi_0, p_0, v)$, then along the geodesic

$$H(\phi_t, p_t) = H(\phi_0, p_0) = \frac{1}{2}\int_X p_0 \cdot v_0 dx = \frac{1}{2}\int_X Av_0 \cdot v_0 dx = \frac{1}{2}\int_X Av_t \cdot v_t dx$$

The stationarity of the Hamiltonian demonstrates the interpretation of the Lagrange multiplier as momentum; integrated against velocity $\dot{\phi}$ gives energy density. The canonical momentum has many names. In optimal control, the flows ϕ is interpreted as the state, and p is interpreted as conjugate state, or conjugate momentum.[92] The geodesi of EL implies specification of the vector fields v_0 or Eulerian momentum Av_0 at $t = 0$, or specification of canonical momentum p_0 determines the flow

The Metric on Geodesic Flows of Landmarks, Surfaces, and Volumes within the Orbit

In Computational anatomy the submanifolds are pointsets, curves, surfaces and subvolumes which are the basic primitive forming the index sets or background space of medically imaged human anatomy. The geodesic flows of the submanifolds such as the landmarks, surface and subvolumes and the distance as measured by the geodesic flows of such coordinates, form the basic measuring and transporting tools of diffeomorphometry.

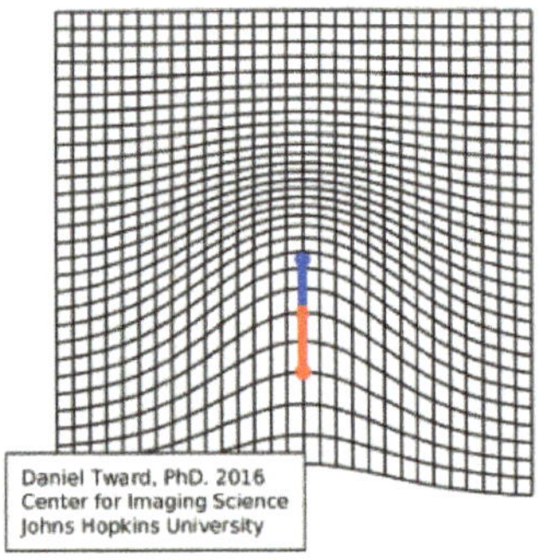

Illustration of geodesic flow for one landmark, demonstrating diffeomorphic motion of background space. Red arrow shows $p_0(1)$, blue curve shows $\varphi_t(x_1)$; black grid shows φ_t.

What is so important about the RKHS norm defining the kinetic energy in the action principle is that the vector fields of the geodesic motions of the submanifolds are superpositions of Green's Kernel's. For landmarks the superposition is a sum of weight kernels weighted by the canonical momentum which determines the inner product, for surfaces it is a surface integral, and for dense volumes it is a volume integral.

At $t = 0$ the geodesic has vector field $v_0 = Kp_0$ determined by the conjugate momentum and the Green's kernel of the inertial operator defining the Eulerian momentum $K = A^{-1}$. The metric distance between coordinate systems connected via the geodesic determined by the induced distance between identity and group element:

$$d_{Diff_V}(id, \varphi) = \|Log_{id}(\varphi)\|_V = \|v_0\|_V = \sqrt{2H(id, p_0)}$$

Landmark and surface submanifolds have Lagrange multiplier associated to a sum and surface integral, respectively; dense volumes an integral with respect to Lebesgue measure.

Landmark or Pointset Geodesics

For Landmarks, the Hamiltonian momentum is defined on the indices, $p(i), i = 1, \ldots, n$ with the inner product given by $(p_t|v_t \circ \phi_t) \doteq \sum_i p_t(i) \cdot v_t \circ \phi_t(x_i)$ and Hamiltonian $H(\phi_t, p_t) = \frac{1}{2} \sum_j \sum_i p_t(i) \cdot K(\phi_t(x_i), \phi_t(x_j))p_t(j)$.

The dynamics take the forms

$$\begin{cases} v_t = \sum_i K(\cdot, \phi_t(x_i))p_t(i), \\ \dot{p}_t(i) = -(Dv_t)^T_{|\phi_t(x_i)} p_t(i), i = 1, 2, ;n \end{cases}$$

with the metric between landmarks $d^2 = (p_0|v_0) = \sum_i p_0(i) \cdot \sum_j K(x_i, x_j)p_0(j)$.

Surface Geodesics

For surfaces, the Hamiltonian momentum is defined across the surface with the inner product

$(p_t \mid v_t \circ \phi_t) \doteq \int_U p_t(u) \cdot v_t \circ \phi_t(m(u))du$, with $H(\phi_t, p_t) = \frac{1}{2}\int_U \int_U p_t(u) \cdot K(\phi_t(m(u)), \phi_t(m(v)))p_t(v)\, du\, dv$

. The dynamics

$$\begin{cases} v_t = \int_U K(\cdot, \phi_t(m(u)))p_t(u)\, du\,, \\ \dot{p}_t(u) = -(Dv_t)^T_{|_{\phi_t(m(u))}} p_t(u), u \in U \end{cases}$$

with the metric between surface coordinates $d^2 = (p_0 \mid v_0) = \int_U p_0(u) \cdot \int_U K(m(u), m(u'))p_0(u')\, du\, du'$

Volume Geodesics

For volumes the Hamiltonian momentum is $(p_t \mid v_t \circ \phi_t) \doteq \int_{\mathbb{R}^3} p_t(x) \cdot v_t \circ \phi_t(x)\, dx$ with $H(\phi_t, p_t) = \frac{1}{2}\int_{\mathbb{R}^3} \int_{\mathbb{R}^3} p_t(x) \cdot K(\phi_t(x), \phi_t(y))p_t(y)\, dx\, dy$. The dynamics $\begin{cases} v_t = \int_X K(\cdot, \phi_t(x))p_t(x)\, dx\,, \\ \dot{p}_t(x) = -(Dv_t)^T_{|_{\phi_t(x)}} p_t(x), x \in \mathbb{R}^3 \end{cases}$

with the metric between volumes $d^2 = (p_0 \mid v_0) = \int_{\mathbb{R}^3} p_0(x) \cdot \int_{\mathbb{R}^3} K(x, y)p_0(y)dy\, dx$.

Conservation laws on Diffeomorphic Shape Momentum for Computational Anatomy

Given the leastaction there is a natural definition of momentum associated to generalized coordinates; the quantity acting against velocity gives energy. The field has studied two forms, the momentum associated to the Eulerian vector field termed Eulerian diffeomorphic shape momentum, and the momentum associated to the initial coordinates or canonical coordinates termed canonical diffeomorphic shape momentum. Each has a conservation law. The conservation of momentum goes hand in hand with the ELGeneral. In Computational anatomy, Av is the Eulerian Momentum since when integrated against Eulerian velocity v gives energy density; operator A the generalized moment of inertia or inertial operator which acting on the Eulerian velocity gives momentum which is conserved along the geodesic:

Eulerian $\quad \frac{d}{dt}\int_X Av_t \cdot ((D\phi_t)w) \circ \phi_t^{-1})dx = 0\,,\ t \in [0,1]$.

Canonical $\quad \frac{d}{dt}\int_X p_t \cdot ((D\phi_t)w)dx = 0\,,\ t \in [0,1]$ for all $w \in V$.

Conservation of Eulerian shape momentum was shown in and follows from **EL-**

General; conservation of canonical momentum was shown in

The proof follow from defining $w_t = ((D\phi_t)w) \circ \phi_t^{-1}$, $\frac{d}{dt} w_t = (Dv_t)w_t - (Dw_t)v_t$ implying

$$\frac{d}{dt}(Av_t|((D\phi_t)w) \circ \phi_t^{-1}) = (\frac{d}{dt}Av_t|((D\phi_t)w) \circ \phi_t^{-1}) + (Av_t|\frac{d}{dt}((D\phi_t)w) \circ \phi_t^{-1}) = (\frac{d}{dt}Av_t|w_t) + (Av_t|(Dv_t)w_t - (Dw_t)v_t) = 0.$$

The proof on Canonical momentum is shown from $\dot{p}_t = -(Dv_t)^T_{|\phi_t} p_t$:

$$\frac{d}{dt}(p_t|(D\phi_t)w) = (\dot{p}_t|(D\phi_t)w) + (p_t|\frac{d}{dt}(D\phi_t)w) = (\dot{p}_t|(D\phi_t)w) + (p_t|(Dv_t)_{|\phi_t}(D\phi_t)w) = 0.$$

Geodesic Interpolation of Information between Coordinate Systems via Variational Problems

Construction of diffeomorphic correspondences between shapes calculates the initial vector field coordinates $v_0 \in V$ and associated weights on the Greens kernels p_0. These initial coordinates are determined by matching of shapes, called **Large Deformation Diffeomorphic Metric Mapping (LDDMM).** LDDMM has been solved for landmarks with and without correspondence and for dense image matchings. curves,[1] surfaces,dense vectorand tensor imagery, and varifolds removing orientation.LDDMM calculates geodesic flows of the **EL-General** onto target coordinates, adding to the action integral $\frac{1}{2}\int_0^1 \int_X Av_t \cdot v_t dx dt$ an endpoint matching condition $E : \phi_1 \to R^+$ measuring the correspondence of elements in the orbit under coordinate system transformation. Existence of solutions were examined for image matching.[22] The solution of the variational problem satisfies the EL-General for $t \in [0, 1)$ with boundary condition.

Matching Based on Minimizing Kinetic Energy Action with Endpoint Condition

$$\min_{\phi:v=\dot{\phi}\circ\phi^{-1},\phi_0=id} C(\phi) \doteq \frac{1}{2}\int_0^1 \int_X Av_t \cdot v_t dx dt + E(\phi_1)$$

$$\begin{cases} \text{Euler Conservation} & \frac{d}{dt}Av_t + ad^*_{v_t}(Av_t) = 0,\ t \in [0,1)\ , \\ \text{Boundary Condition} & \phi_0 = id,\ Av_1 = -\frac{\partial E(\phi)}{\partial \phi}|_{\phi=\phi_1}\ . \end{cases}$$

Conservation from EL-General extends the B.C. at t=1 to the rest of the path $t \in [0, 1)$ The inexact matching problem with the endpoint matching term $E(\phi_1)$ has several alternative forms. One of the key ideas of the stationarity of the Hamiltonian along the geodesic solution is the integrated running cost reduces to initial cost at t=0, geodesics of the EL-General are determined by their initial condition v_0.

Dense Image Matching in Computational Anatomy

Dense image matching has a long history now with the earliest efforts exploiting a small deformation framework. Large deformations began in the early 90's, with the first existence to solutions to the variational problem for flows of diffeomorphisms for dense image matching established in. Beg solved via one of the earliest LDDMM algorithms based on solving the variational matching with endpoint defined by the dense imagery with respect to the vector fields, taking variations with respect to the vector fields. Another solution for for dense image matching reparameterizes the optimization problem in terms of the state $q_t \doteq I \circ \phi_t^{-1}, q_0 = I$ giving the solution in terms of the infinitesimal action defined by the advection equation.

LDDMM Dense Image Matching

For Beg's LDDMM, denote the Image $I(x), x \in X$ with group action $\phi \cdot I \doteq I \circ \phi^{-1}$. Viewing this as an optimal control problem, the state of the system is the diffeomorphic flow of coordinates $\phi_t, t \in [0,1]$ with the dynamics relating the control $v_t, t \in [0,1]$ to the state given by $\dot{\phi} = v \circ \phi$. The endpoint matching condition $E(\phi_1) \doteq \|I \circ \phi_1^{-1} - I'\|^2$

gives the variational problem

$$\min_{v:\dot{\phi}=v\circ\phi} C(v) \doteq \frac{1}{2}\int_0^1 \int_X Av_t \cdot v_t dx dt + \frac{1}{2}\int_{\mathbb{R}^3} |I \circ \phi_1^{-1}(x) - I'(x)|^2 dx$$

$$\begin{cases} \text{Endpoint Condition:} & Av_1 = \mu_1 dx, \mu_1 = (I \circ \phi_1^{-1} - I')\nabla(I \circ \phi_1^{-1}), \\ \text{Conservation:} & Av_t = \mu_t \, dx, \; \mu_t = (D\phi_t^{-1})^T \mu_0 \circ \phi_t^{-1} |D\phi_t^{-1}|. \\ & \mu_0 = (I - I' \circ \phi_1)\nabla I |D\phi_1|. \end{cases}$$

Beg's iterative LDDMM algorithm has fixed points which satisfy the necessary optimizer conditions. The iterative algorithm is given in Beg's LDDMM algorithm for dense image matching.

Hamiltonian LDDMM in the Reduced Advected State

Denote the Image $I(x), x \in X$, with state $q_t \doteq I \circ \phi_t^{-1}$ and the dynamics related state and control given by the advective term $\dot{q}_t = -\nabla q_t \cdot v_t$. The endpoint $E(q_1) \doteq \|q_1 - I'\|^2$ gives the variational problem

$$\min_{q:\dot{q}=v\circ q} C(v) \doteq \frac{1}{2}\int_0^1 \int_X Av_t \cdot v_t dx dt + \frac{1}{2}\int_{\mathbb{R}^3} |q_1(x) - I'(x)|^2 dx$$

Diffusion Tensor Image Matching in Computational Anatomy

Dense LDDMM tensor matching solves the variational problem matching between coordinate system based on the principle eigenvectors of the diffusion tensor MRI image (DTI) denoted $M(x), x \in \mathbb{R}^3$ consisting of the 3x3 tensor at every voxel. Several of the group actions defined based on the Frobenius matrix norm between square matrices

$\|A\|_F^2 \doteq trace A^T A$. Shown in the accompanying figure is a DTI image illustrated via its color map depicting the eigenvector orientations of the DTI matrix at each voxel with color determined by the orientation of the directions. Denote the 3x3 tensor image $M(x), x \in \mathbb{R}^3$ with eigen-elements $\{\lambda_i(x), e_i(x), i = 1,2,3\}$ $\lambda_1 \geq \lambda_2 \geq \lambda_3$.

Coordinate system transformation based on DTI imaging has exploited two actions one based on the principle eigen-vector or entire matrix.

LDDMM matching based on the principal eigenvector of the diffusion tensor matrix takes the image $I(x), x \in \mathbb{R}^3$ as a unit vector field defined by the first eigenvector. The group action becomes

$$\varphi \cdot I = \begin{cases} \dfrac{D_{\varphi^{-1}} \varphi I \circ \varphi^{-1} \| I \circ \varphi^{-1} \|}{\| D_{\varphi^{-1}} \varphi I \circ \varphi^{-1} \|} & I \circ \varphi \neq 0; \\ 0 & \textbf{otherwise.} \end{cases}$$

LDDMM matching based on the entire tensor matrix has group action becomes $\varphi \cdot M = (\lambda_1 \hat{e}_1 \hat{e}_1^T + \lambda_2 \hat{e}_2 \hat{e}_2^T + \lambda_3 \hat{e}_3 \hat{e}_3^T) \circ \varphi^{-1}$, transformed eigenvectors

$$\hat{e}_1 = \frac{D\varphi e_1}{\|D\varphi e_1\|}, \quad \hat{e}_2 = \frac{D\varphi e_2 - \langle \hat{e}_1, D\varphi e_2 \rangle \hat{e}_1}{\sqrt{\|D\varphi e_2\|^2 - \langle \hat{e}_1, D\varphi e_2 \rangle^2}}, \quad \hat{e}_3 = \hat{e}_1 \times \hat{e}_2.$$

The variational problem matching onto the principal eigenvector or the matrix is described LDDMM Tensor Image Matching.

Metamorphosis

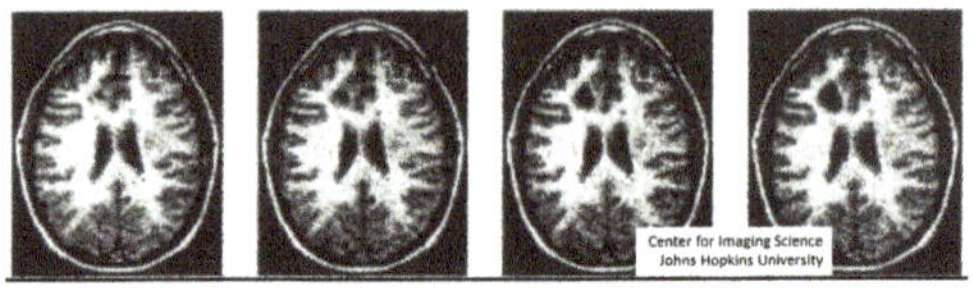

Demonstrating metamorphosis allowing both diffeomorphic change in coordinate transformation as well as change in image intensity as associated to early Morphing technologies such as the Michael Jackson video. Notice the insertion of tumor gray level intensity which does not exist in template.

The principle mode of variation represented by the orbit model is change of coordinates. For setting in which pairs of images are not related by diffeomorphisms but have photometric variation or image variation not represented by the template, active appearance modelling has been introduced, originally by Edwards-Cootes-Taylor and in 3D medical imaging in. In the context of Computational Anatomy in which metrics on the anatomical orbit has been studied, **metamorphosis** for modelling structures such as tumors and photometric changes which are not resident in the template was introduced in for Magnetic Resonance image models, with many subsequent developments extending the metamorphosis framework.

For image matching the image metamorphosis framework enlarges the action so that $t \mapsto (\phi_t, I_t)$ with action $\phi_t \cdot I_t \doteq I_t \circ \phi_t^{-1}$ In this setting metamorphosis combines both the diffeomorphic coordinate system transformation of Computational Anatomy as well as the early morphing technologies which only faded or modified the photometric or image intensity alone.

Then the matching problem takes a form with equality boundary conditions:

$$\min_{(v,I)} \frac{1}{2} \int_0^1 \left((Av_t \mid v_t) + \|\dot{I}_t \circ \phi_t^{-1}\|^2/\sigma^2 \right) dt \text{ subject to } \phi_0 = id, I_0 = \text{fixed}, I_1 = \text{fixed}$$

Matching Landmarks, Curves, Surfaces

Transforming coordinate systems based on Landmark point or fiducial marker features dates back to Bookstein's early work on small deformation spline methods for interpolating correspondences defined by fiducial points to the two-dimensional or three-dimensional background space in which the fiducials are defined. Large deformation landmark methods came on in the late 90's. The above Figure depicts a series of landmarks associated three brain structures, the amygdala, entorhinal cortex, and hippocampus.

Matching geometrical objects like unlabelled point distributions, curves or surfaces is another common problem in Computational Anatomy. Even in the discrete setting where these are commonly given as vertices with meshes, there are no predetermined correspondences between points as opposed to the situation of landmarks described above. From the theoretical point of view, while any submanifold X in $\mathbb{R}^3$, $d = 1, 2, 3$ can be parameterized in local charts $m : u \in U \subset \mathbb{R}^{0,1,2,3} \to \mathbb{R}^3$ all reparametrizations of these charts give geometrically the same manifold. Therefore, early on in Computational anatomy, investigators have identified the necessity of parametrization invariant representations. One indispensable requirement is that the end-point matching term between two submanifolds is itself independent of their parametrizations. This can be achieved via concepts and methods borrowed from Geometric measure theory, in particular currents and varifolds which have been used extensively for curve and surface matching.

Landmark or Point Matching with Correspondence

Denoted the landmarked shape $X \doteq \{x_1, \ldots, x_n\} \subset \mathbb{R}^3$ with endpoint $E(\phi_1) \doteq \sum_i \|\phi_1(x_i) - x_i'\|^2$ the variational problem becomes

$$\min_{\phi: v = \dot{\phi} \circ \phi^{-1}} C(\phi) \doteq \frac{1}{2} \int (Av_t | v_t) dt + \frac{1}{2} \sum_i \|\phi_1(x_i) - x_i'\|^2$$

The geodesic Eulerian momentum is a generalized function $Av_t \in V^*, t \in [0, 1]$ supported on the landmarked set in the variational problem.The endpoint condition with conservation implies the initial momentum at the identiy of the group:

$$\begin{cases} \text{Endpoint Condition:} & Av_1 = \sum_{i=1}^{n} p_1(i)\delta_{\phi_1(x_i)}, p_1(i) = (x'_i - \phi_1(x_i)), \\ \text{Conservation:} & Av_t = \sum_{i=1}^{n} p_t(i)\delta_{\phi_t(x_i)}, \; p_t(i) = (D\phi_{t1})^T_{|\phi_t(x_i)} p_1(i), \; \phi_{t1} \doteq \phi_1 \circ \phi_t^{-1}, \\ & Av_0 = \sum_i \delta_{x_i}(\cdot) p_0(i) \text{ with } p_0(i) = (D\phi_1)^T_{|x_i}(x'_i - \phi_1(x_i)) \end{cases}$$

The iterative algorithm for large deformation diffeomorphic metric mapping for landmarks is given.

Measure Matching: Unregistered Landmarks

Glaunes and co-workers first introduced diffeomorphic matching of pointsets in the general setting of matching distributions. As opposed to landmarks, this includes in particular the situation of weighted point clouds with no predefined correspondences and possibly different cardinalities. The template and target discrete point clouds are represented as two weighted cardinalities. The template and target discrete point clouds are represented as two weighted sums of Diracs $\mu_m = \sum_{i=1}^{n} \rho_i \delta_{x_i}$ and $\mu_{m'} = \sum_{i=1}^{n'} \rho'_i \delta_{x'_i}$ living in the space of signed measures of R^3 The space is equipped with a Hilbert metric obtained from a real positive kernel $k(x, y)$ on $\mathbb{R}^3$ giving the following norm:

$$\|\mu_m\|^2_{\text{mea}} = \sum_{i,j=1}^{n} \rho_i \rho_j k(x_i, x_j)$$

The matching problem between a template and target point cloud may be then formulated using this kernel metric for the endpoint matching term:

$$\min_{\phi: v = \dot{\phi} \circ \phi^{-1}} C(\phi) \doteq \frac{1}{2} \int (Av_t \mid v_t)\, dt + \frac{1}{2} \|\mu_{\phi_1 \cdot m} - \mu_{m'}\|^2_{\text{mea}}$$

where $\mu_{\phi_1 \cdot m} = \sum_{i=1}^{n} \rho_i \delta_{\phi_1(x_i)}$ is the distribution transported by the deformation.

Curve Matching

In the one dimensional case, a curve in 3D can be represented by an embedding $m : u \in [0,1] \to \mathbb{R}^3$ and the group action of *Diff* becomes $\phi \cdot m = \phi \circ m$. However, the correspondence between curves and embeddings is not one to one as the any reparametrization $m \circ \gamma$, for γ a diffeomorphism of the interval [0,1], represents geometrically the same curve. In order to preserve this invariance in the end-point matching term, several extensions of the previous 0-dimensional measure matching approach can be considered.

Curve matching with currents

In the situation of oriented curves, currents give an efficient setting to construct invariant matching terms. In such representation, curves are interpreted as elements of a functional space dual to the space vector fields, and compared through kernel norms on these spaces. Matching of two curves m and m' writes eventually as the variational problem

$$\min_{\phi:v=\dot{\phi}\circ\phi^{-1}} C(\phi) \doteq \frac{1}{2}\int (Av_t \mid v_t)\,dt + \frac{1}{2}\|\mathcal{C}_{\phi_1\cdot m} - \mathcal{C}_{m'}\|^2_{\text{cur}}$$

with the endpoint term $E(\phi_1) = \|\mathcal{C}_{\phi_1\cdot m} - \mathcal{C}_{m'}\|^2_{\text{cur}}/2$ is obtained from the norm

$$\|\mathcal{C}_m\|^2_{\text{cur}} = \int_0^1\int_0^1 K_C(m(u), m(v))\partial m(u)\cdot\partial m(v)\,du\,dv$$

the derivative $\partial m(u)$ being the tangent vector to the curve and K_C a given matrix kernel of R³ Such expressions are invariant to any positive reparametrizations of m and m' and thus still depend on the orientation of the two curves.

- Curve matching with varifolds

Varifold is an alternative to currents when orientation becomes an issue as for instance in situations involving multiple bundles of curves for which no "consistent" orientation may be defined. Varifolds directly extend 0-dimensional measures by adding an extra tangent space direction to the position of points, leading to represent curves as measures on the product of $\mathbb{R}^3$ and the Grassmannian of all straight lines in $\mathbb{R}^3$ The matching problem between two curves then consists in replacing the endpoint matching term by $E(\phi_1) = \|\mathcal{V}_{\phi_1\cdot m} - \mathcal{V}_{m'}\|^2_{cur}/2$

$$\|\mathcal{V}_m\|^2_{var} = \int_0^1\int_0^1 k_{\mathbb{R}^3}(m(u), m(v))k_{\text{Gr}}([\partial m(u)], [\partial m(v)])\,|\partial m(u)||\partial m(v)|\,du\,dv$$

Where $[\partial m(u)]$ is the non-oriented line directed by tangent vector $\partial m(u)$ and $k_{\mathbb{R}^3}, k_{\text{Gr}}$ two scalar kernels respectively on $\mathbb{R}^3$ and the Grassmannian. Due to the inherent non-oriented nature of the Grassmannian representation, such expressions are invariant to positive and negative reparametrizations.

Surface Matching

Surface matching share many similarities with the case of curves. Surfaces in $\mathbb{R}^3$ are parametrized in local charts by embeddings $m : u \in U \subset \mathbb{R}^2 \to \mathbb{R}^3$ with all reparametrizations $m \circ \gamma$ with γ a diffeomorphism of U being equivalent geometrically. Currents and varifolds can be also used to formalize surface matching.

- Surface matching with currents

Oriented surfaces can be represented as 2-currents which are dual to differential 2-forms.

In $\mathbb{R}^3$ one can further identify 2-forms with vector fields through the standard wedge product of 3D vectors. In that setting, surface matching writes again:

$$\min_{\phi:v=\dot{\phi}\circ\phi^{-1}} C(\phi) \doteq \frac{1}{2}\int (Av_t \mid v_t)\,dt + \frac{1}{2}\|\mathcal{C}_{\phi_1\cdot m} - \mathcal{C}_{m'}\|_{\text{cur}}^2$$

with the endpoint term $E(\phi_1) = \|\mathcal{C}_{\phi_1\cdot m} - \mathcal{C}_{m'}\|_{\text{cur}}^2/2$ given through the norm

$$\|\mathcal{C}_m\|_{\text{cur}}^2 = \iint_{U\times U} K_C(m(u), m(v))\vec{n}(u)\cdot\vec{n}(v)\,du\,dv$$

with $\vec{n} = \partial_{u_1} m \wedge \partial_{u_2} m$ the normal vector to the surface parametrized by m.

This surface mapping algorithm has been validated for brain cortical surfaces against CARET and FreeSurfer.[122] LDDMM mapping for multiscale surfaces is discussed in.[123]

- Surface matching with varifolds

For non-orientable or non-oriented surfaces, the varifold framework is often more adequate. Identifying the parametric surface m with a varifold $\mathcal{V}_m$ in the space of measures on the product of $\mathbb{R}^3$ and the Grassmannian, one simply replaces the previous current metric $\|\mathcal{C}_m\|_{\text{cur}}^2$ by:

$$\|\mathcal{V}_m\|_{\text{var}}^2 = \iint_{U\times U} k_{\mathbb{R}^3}(m(u), m(v))k_{\text{Gr}}\left([\vec{n}(u)], [\vec{n}(v)]\right)|\vec{n}(u)||\vec{n}(v)|\,du\,dv$$

Where $[\vec{n}(u)]$ is the (non-oriented) line directed by the normal vector to the surface.

Growth and Atrophy from Longitudinal Time-Series

There are many settings in which there are a series of measurements, a time-series to which the underlying coordinate systems will be matched and flowed onto. This occurs for example in the dynamic growth and atrophy models and motion tracking such as have been explored in An observed time sequence is given and the goal is to infer the time flow of geometric change of coordinates carrying the exemplars or templars through the period of observations.

The generic time-series matching problem considers the series of times is $0 < t_1 < \ldots t_K = 1$. The flow optimizes at the series of costs $E(t_k), k = 1, \ldots, K$ giving optimization problems of the form

$$\min_{\phi: v=\dot{\phi}\circ\phi^{-1}, \phi_0=id} C(\phi) \doteq \frac{1}{2}\int_0^1 (Av_t \mid v_t)\, dt + \sum_{k=1}^{K} E(\phi_{t_k})$$

There have been at least three solutions offered thus far, piecewise geodesic, principal geodesic and splines.

The Random Orbit Model of Computational Anatomy

The **random orbit model** of Computational Anatomy first appeared in modelling the change in coordinates associated to the randomness of the group acting on the templates, which induces the randomness on the source of images in the anatomical orbit of shapes and forms and resulting observations through the medical imaging devices. Such a **random orbit model** in which randomness on the group induces randomness on the images was examined for the Special Euclidean Group for object recognition in.

Depicted in the figure is a depiction of the random orbits around each exemplar, $m_0 \in \mathcal{M}$, generated by randomizing the flow by generating the initial tangent space vector field at the identity $v_0 \in V$ and then generating random object $n \doteq Exp_{id}(v_0) \cdot m_0 \in \mathcal{M}$.

The random orbit model induces the prior on shapes and images $I \in \mathcal{I}$ conditioned on a particular atlas $I_a \in \mathcal{I}$ For this the generative model generates the mean field I as a random change in coordinates of the template according to $I \doteq \phi \cdot I_a$ where the diffeomorphic change in coordinates is generated randomly via the geodesic flows. The prior on random transformations $\pi_{Diff}(d\phi)$ on $Diff_V$ is induced by the flow $Exp_{id}(v)$, with $v \in V$ constructed as a Gaussian random field prior $\pi_V(dv)$ The density on the random observables at the output of the sensor $I^D \in \mathcal{I}^D$ are given by

$$p(I^D|I_a) = \int_V p(I^D|Exp_{id}(v) \cdot I_a)\pi_V(dv) \, .$$

Shown in the Figure on the right the cartoon orbit, are a random spray of the subcortical manifolds generated by randomizing the vector fields v_0 supported over the submanifolds.

The Bayesian Model of Computational Anatomy

The central statistical model of Computational Anatomy in the context of medical imaging has been the source-channel model of Shannon theory; the source is the deformable template of images $I \in \mathcal{I}$, the channel outputs are the imaging sensors with observables $I^D \in \mathcal{I}^D$

See The Bayesian model of computational anatomy for discussions (i) MAP estimation with multiple atlases, (ii) MAP segmentation with multiple atlases, MAP estimation of templates from populations.

Statistical Shape Theory in Computational Anatomy

Shape in computational anatomy is a local theory, indexing shapes and structures to templates to which they are bijectively mapped. Statistical shape in Computational Anatomy is the empirical study of diffeomorphic correspondences between populations and common template coordinate systems. Interestingly, this is a strong departure from Procrustes Analyses and shape theories pioneered by David G. Kendall in that the central group of Kendall's theories are the finite-dimensional Lie groups, whereas the theories of shape in Computational Anatomy have focused on the diffeomorphism group, which to first order via the Jacobian can be thought of as a field–thus infinite dimensional–of low-dimensional Lie groups of scale and rotations.

The random orbit model provides the natural setting to understand empirical shape and shape statistics within Computational anatomy since the non-linearity of the induced probability law on anatomical shapes and forms $m \in \mathcal{M}$ is induced via the reduction to the vector fields $v_0 \in V$ at the tangent space at the identity of the diffeomorphism group. The successive flow of the Euler equation induces the random space of shapes and forms $Exp_{id}(v_0) \cdot m \in \mathcal{M}$.

Performing empirical statistics on this tangent space at the identity is the natural way for inducing probability laws on the statistics of shape. Since both the vector fields and the Eulerian momentum Av_0 are in a Hilbert space the natural model is one of a Gaussian random field, so that given test function $w \in V$ then the inner-products with the test functions are Gaussian distributed with mean and covariance.

This is depicted in the accompanying figure where sub-cortical brain structures are depicted in a two-dimensional coordinate system based on inner products of their initial vector fields that generate them from the template is shown in a 2-dimensional span of the Hilbert space.

Template Estimation from Populations

The study of shape and statistics in populations are local theories, indexing shapes and structures to templates to which they are bijectively mapped. Statistical shape is then the study of diffeomorphic correspondences relative to the template. A core operation is the generation of templates from populations, estimating a shape that is matched to the population. There are several important methods for generating templates including methods based on Frechet averaging, and statistical approaches based on the expectation-maximization algorithm and the Bayes Random orbit models of Computational anatomy. Shown in the accompanying figure is a subcortical template reconstruction from the population of MRI subjects.

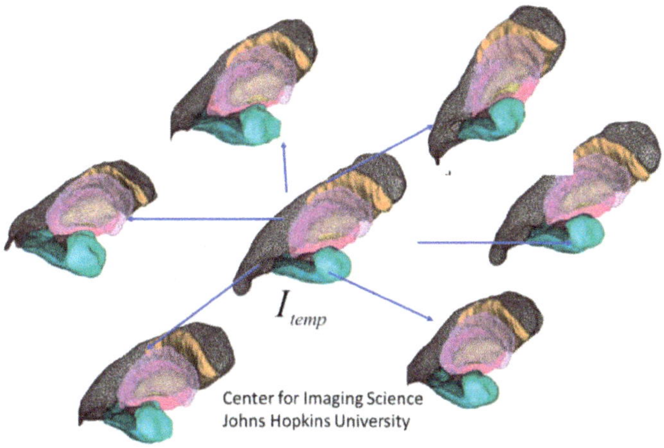

Depicting template estimation from multiplie subcortical surfaces in populations of MR images using the EM-algorithm solution of Ma.

CONNECTIONISM

Connectionism is a set of approaches in the fields of artificial intelligence, cognitive psychology, cognitive science, neuroscience, and philosophy of mind, that models mental or behavioral phenomena as the emergent processes of *interconnected networks of simple units*. The term was introduced by Donald Hebb in the 1940s. There are many forms of connectionism, but the most common forms use neural network models.

Basic Principles

The central connectionist principle is that mental phenomena can be described by interconnected networks of simple and often uniform units. The form of the connections and the units can vary from model to model. For example, units in the network could represent neurons and the connections could represent synapses like in the brain of a human being.

Spreading Activation

In most connectionist models, networks change over time. A closely related and very common aspect of connectionist models is *activation*. At any time, a unit in the network has an activation, which is a numerical value intended to represent some aspect of the unit. For example, if the units in the model are neurons, the activation could represent the probability that the neuron would generate an action potential spike. Activation typically spreads to all the other units connected to it. Spreading activation is always a feature of neural network models, and it is very common in connectionist models used by cognitive psychologists.

Neural Networks

Neural networks are by far the most commonly used connectionist model today. Though there are a large variety of neural network models, they almost always follow two basic principles regarding the mind:

- Any mental state can be described as an (N)-dimensional vector of numeric activation values over neural units in a network.
- Memory is created by modifying the strength of the connections between neural units. The connection strengths, or "weights", are generally represented as an N×N matrix.

Most of the variety among neural network models comes from:

- *Interpretation of units*: Units can be interpreted as neurons or groups of neurons.
- *Definition of activation*: Activation can be defined in a variety of ways. For example, in a Boltzmann machine, the activation is interpreted as the probability of generating an action potential spike, and is determined via a logistic function on the sum of the inputs to a unit.
- *Learning algorithm*: Different networks modify their connections differently. In general, any mathematically defined change in connection weights over time is referred to as the "learning algorithm".

Connectionists are in agreement that recurrent neural networks (directed networks wherein connections of the network can form a directed cycle) are a better model of the brain than feedforward neural networks (directed networks with no cycles, called DAG). Many recurrent connectionist models also incorporate dynamical systems theory. Many researchers, such as the connectionist Paul Smolensky, have argued that connectionist models will evolve toward fully continuous, high-dimensional, non-linear, dynamic systems approaches.

Biological Realism

The neural network branch of connectionism suggests that the study of mental activity is really the study of neural systems. This links connectionism to neuroscience, and models involve varying degrees of biological realism. Connectionist work in general need not be biologically realistic, but some neural network researchers, computational neuroscientists, try to model the biological aspects of natural neural systems very closely in so-called "neuromorphic networks". Many authors find the clear link between neural activity and cognition to be an appealing aspect of connectionism.

Learning

The weights in a neural network are adjusted according to some learning rule or algorithm, such as Hebbian learning. Thus, connectionists have created many sophisticated learning procedures for neural networks. Learning always involves modifying the connection weights. In general, these involve mathematical formulas to determine the change in weights when given sets of data consisting of activation vectors for some subset of the neural units.

By formalizing learning in such a way, connectionists have many tools. A very common strategy in connectionist learning methods is to incorporate gradient descent over an error surface in a space defined by the weight matrix. All gradient descent learning in connectionist models involves changing each weight by the partial derivative of the error surface with respect to the weight. Backpropagation (BP), first made popular in the 1980s, is probably the most commonly known connectionist gradient descent algorithm today.

Connectionism can be traced to ideas more than a century old, which were little more than speculation until the mid-to-late 20th century. Through his work on the structure of the nervous system for which he won the Nobel Prize in 1906, the Spanish Santiago Ramón y Cajal established the basis for studies of neural networks, but it wasn't until the 1980s that connectionism became a popular perspective among scientists.

Parallel Distributed Processing (PDP)

The prevailing connectionist approach today was originally known as parallel distributed processing (PDP). It was an artificial neural network approach that stressed the parallel nature of neural processing, and the distributed nature of neural representations. It provided a general mathematical framework for researchers to operate in. The framework involved eight major aspects:

- A set of *processing units*, represented by a set of integers.
- An *activation* for each unit, represented by a vector of time-dependent functions.
- An *output function* for each unit, represented by a vector of functions on the activations.
- A *pattern of connectivity* among units, represented by a matrix of real numbers indicating connection strength.
- A *propagation rule* spreading the activations via the connections, represented by a function on the output of the units.
- An *activation rule* for combining inputs to a unit to determine its new activation, represented by a function on the current activation and propagation.
- A *learning rule* for modifying connections based on experience, represented by a change in the weights based on any number of variables.
- An *environment* that provides the system with experience, represented by sets of activation vectors for some subset of the units.

A lot of the research that led to the development of PDP was done in the 1970s, but PDP became popular in the 1980s with the release of the books *Parallel Distributed Processing: Explorations in the Microstructure of Cognition - Volume 1 (foundations)* and *Volume 2 (Psychological and Biological Models)*, by James L. McClelland, David E. Rumelhart and the PDP Research Group. The books are now considered seminal connectionist works, and it is now common to fully equate PDP and connectionism, although the term "connectionism" is not used in the books.

Earlier Work

PDP's direct roots were the perceptron theories of researchers such as Frank Rosenblatt from the 1950s and 1960s. But perceptron models were made very unpopular by the book *Perceptrons* by Marvin Minsky and Seymour Papert, published in 1969. It demonstrated the limits on the sorts of functions that single-layered (no hidden layer) perceptrons can calculate, showing that even simple functions like the exclusive disjunction (XOR) could not be handled properly. The PDP books overcame this limitation by showing that multi-level, non-linear neural networks were far more robust and could be used for a vast array of functions.

Many earlier researchers advocated connectionist style models, for example in the 1940s and 1950s, Warren McCulloch and Walter Pitts (MP neuron), Donald Olding Hebb, and Karl Lashley. McCulloch and Pitts showed how neural systems could implement first-order logic: Their classic paper «A Logical Calculus of Ideas Immanent in Nervous Activity» (1943) is important in this development here. They were influenced by the important work of Nicolas Rashevsky in the 1930s. Hebb contributed greatly to speculations about neural functioning, and proposed a learning principle, Hebbian learning, that is still used today. Lashley argued for distributed representations as a result of his failure to find anything like a localized engram in years of lesion experiments.

Connectionism Apart from PDP

Though PDP is the dominant form of connectionism, other theoretical work should also be classified as connectionist.

Many connectionist principles can be traced to early work in psychology, such as that of William James. Psychological theories based on knowledge about the human brain were fashionable in the late 19th century. As early as 1869, the neurologist John Hughlings Jackson argued for multi-level, distributed systems. Following from this lead, Herbert Spencer's *Principles of Psychology*, 3rd edition (1872), and Sigmund Freud's *Project for a Scientific Psychology* (composed 1895) propounded connectionist or proto-connectionist theories. These tended to be speculative theories. But by the early 20th century, Edward Thorndike was experimenting on learning that posited a connectionist type network.

In the 1950s, Friedrich Hayek proposed that spontaneous order in the brain arose out of decentralized networks of simple units. Hayek›s work was rarely cited in the PDP literature until recently.

Another form of connectionist model was the relational network framework developed by the linguist Sydney Lamb in the 1960s. Relational networks have been only used by linguists, and were never unified with the PDP approach. As a result, they are now used by very few researchers.

There are also hybrid connectionist models, mostly mixing symbolic representations with neural network models. The hybrid approach has been advocated by some researchers (such as Ron Sun).

Connectionism Vs. Computationalism Debate

As connectionism became increasingly popular in the late 1980s, some researchers (including Jerry Fodor, Steven Pinker and others) reacted against it. They argued that connectionism, as then developing, threatened to obliterate what they saw as the progress being made in the fields of cognitive science and psychology by the classical approach of computationalism. Computationalism is a specific form of cognitivism that argues that mental activity is computational, that is, that the mind operates by performing purely formal operations on symbols, like a Turing machine. Some researchers argued that the trend in connectionism represented a reversion toward associationism and the abandonment of the idea of a language of thought, something they saw as mistaken. In contrast, those very tendencies made connectionism attractive for other researchers.

Connectionism and computationalism need not be at odds, but the debate in the late 1980s and early 1990s led to opposition between the two approaches. Throughout the debate, some researchers have argued that connectionism and computationalism are fully compatible, though full consensus on this issue has not been reached. Differences between the two approaches include the following:

- Computationalists posit symbolic models that are structurally similar to underlying brain structure, whereas connectionists engage in "low-level" modeling, trying to ensure that their models resemble neurological structures.
- Computationalists in general focus on the structure of explicit symbols (mental models) and syntactical rules for their internal manipulation, whereas connectionists focus on learning from environmental stimuli and storing this information in a form of connections between neurons.
- Computationalists believe that internal mental activity consists of manipulation of explicit symbols, whereas connectionists believe that the manipulation of explicit symbols provides a poor model of mental activity.
- Computationalists often posit domain specific symbolic sub-systems designed to support learning in specific areas of cognition (e.g., language, intentionality, number), whereas connectionists posit one or a small set of very general learning-mechanisms.

Despite these differences, some theorists have proposed that the connectionist architecture is simply the manner in which organic brains happen to implement the symbol-manipulation system. This is logically possible, as it is well known that connectionist models can implement symbol-manipulation systems of the kind used in computationalist models, as indeed they must be able if they are to explain the human ability to perform symbol-manipulation tasks. But the debate rests on whether this symbol manipulation forms the foundation of cognition in general, so this is not a potential vindication of computationalism. Nonetheless, computational descriptions may be helpful high-level descriptions of cognition of logic, for example.

The debate largely centred on logical arguments about whether connectionist networks could produce the syntactic structure observed in this sort of reasoning. This was later achieved, although using processes unlikely to be possible in the brain, thus the debate persisted. As of 2016, progress in neurophysiology and general advances in the understanding of neural networks have led to the successful modelling of a great many of these early problems, and the debate about fundamental cognition has, thus, largely been decided among neuroscientists in favour of connectionism. However, these fairly recent developments have yet to reach consensus acceptance among those working in other fields, such as psychology or philosophy of mind.

Part of the appeal of computational descriptions is that they are relatively easy to interpret, and thus may be seen as contributing to our understanding of particular mental processes, whereas connectionist models are in general more opaque, to the extent that they may be describable only in very general terms (such as specifying the learning algorithm, the number of units, etc.), or in unhelpfully low-level terms. In this sense connectionist models may instantiate, and thereby provide evidence for, a broad theory of cognition (i.e., connectionism), without representing a helpful theory of the particular process that is being modelled. In this sense the debate might be considered as to some extent reflecting a mere difference in the level of analysis in which particular theories are framed.

The recent popularity of dynamical systems in philosophy of mind have added a new perspective on the debate; some authors now argue that any split between connectionism and computationalism is more conclusively characterized as a split between computationalism and dynamical systems.

MEDICAL IMAGE COMPUTING

Medical image computing (MIC) is an interdisciplinary field at the intersection of computer science, data science, electrical engineering, physics, mathematics and medicine. This field develops computational and mathematical methods for solving problems pertaining to medical images and their use for biomedical research and clinical care.

The main goal of MIC is to extract clinically relevant information or knowledge from medical images. While closely related to the field of medical imaging, MIC focuses on the computational analysis of the images, not their acquisition. The methods can be grouped into several broad categories: image segmentation, image registration, image-based physiological modeling, and others.

Data Forms

Medical image computing typically operates on uniformly sampled data with regular x-y-z spatial spacing (images in 2D and volumes in 3D, generically referred to as images). At each sample point, data is commonly represented in integral form such as signed and unsigned short (16-bit), although forms from unsigned char (8-bit) to 32-bit float are not uncommon. The particular meaning of the data at the sample point depends on modality: for example a CT acquisition collects radiodensity values, while a MRI acquisition may collect T1 or T2-weighted images. Longitudinal, time-varying acquisitions may or may not acquire images with regular time steps. Fan-like images due to modalities such as curved-array ultrasound are also common and require different representational and algorithmic techniques to process. Other data forms include sheared images due to gantry tilt during acquisition; and unstructured meshes, such as hexahedral and tetrahedral forms, which are used in advanced biomechanical analysis (e.g., tissue deformation, vascular transport, bone implants).

Segmentation

Segmentation is the process of partitioning an image into different segments. In medical imaging, these segments often correspond to different tissue classes, organs, pathologies, or other biologically relevant structures. Medical image segmentation is made difficult by low contrast, noise, and other imaging ambiguities. Although there are many computer vision techniques for image segmentation, some have been adapted specifically for medical image computing. Below is a sampling of techniques within this field; the implementation relies on the expertise that clinicians can provide.

- **Atlas-Based Segmentation**: For many applications, a clinical expert can manually label several images; segmenting unseen images is a matter of extrapolating from these manually labeled training images. Methods of this style are typically referred to as atlas-based segmentation methods. Parametric atlas methods typically combine these training images into a single atlas image, while nonparametric atlas methods typically use all of the training images separately. Atlas-based methods usually require the use of image registration in order to align the atlas image or images to a new, unseen image.

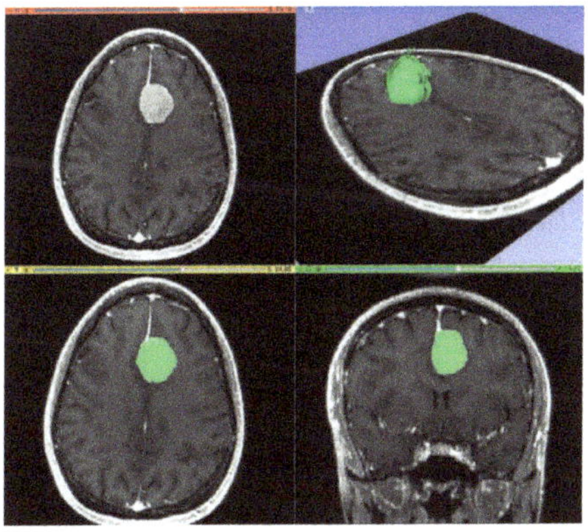

A T1 weighted MR image of the brain of a patient with a meningioma after injection of a MRI contrast agent (top left), and the same image with the result of an interactive segmentation overlaid in green (3D model of the segmentation on the top right, axial and coronal views at the bottom).

- **Shape-Based Segmentation**: Many methods parametrize a template shape for a given structure, often relying on control points along the boundary. The entire shape is then deformed to match a new image. Two of the most common shape-based techniques are Active Shape Models and Active Appearance Models. These methods have been very influential, and have given rise to similar models.
- **Image-Based segmentation**: Some methods initiate a template and refine its shape according to the image data while minimizing integral error measures, like the Active contour model and its variations.

- **Interactive Segmentation**: Interactive methods are useful when clinicians can provide some information, such as a seed region or rough outline of the region to segment. An algorithm can then iteratively refine such a segmentation, with or without guidance from the clinician. Manual segmentation, using tools such as a paint brush to explicitly define the tissue class of each pixel, remains the gold standard for many imaging applications. Recently, principles from feedback control theory have been incorporated into segmentation, which give the user much greater flexibility and allow for the automatic correction of errors.

However, there are some other classification of image segmentation methods which are similar to above categories. Moreover, we can classify another group as "Hybrid" which is based on combination of methods.

Registration

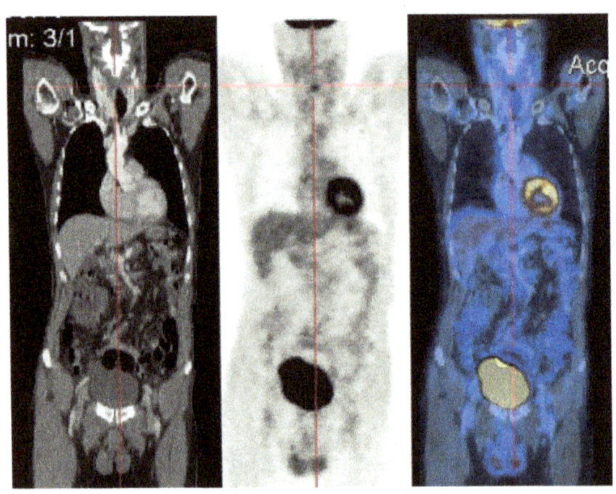

CT image (left), PET image (center) and overlay of both (right) after correct registration.

Image registration is a process that searches for the correct alignment of images. In the simplest case, two images are aligned. Typically, one image is treated as the target image and the other is treated as a source image; the source image is transformed to match the target image. The optimization procedure updates the transformation of the source image based on a similarity value that evaluates the current quality of the alignment. This iterative procedure is repeated until a (local) optimum is found. An example is the registration of CT and PET images to combine structural and metabolic information (see figure).

Image registration is used in a variety of medical applications:

- Studying temporal changes. Longitudinal studies acquire images over several months or years to study long-term processes, such as disease progression. Time series correspond to images acquired within the same session (seconds or minutes). They can be used to study cognitive processes, heart deformations and respiration.

- Combining complementary information from different imaging modalities. An example is the fusion of anatomical and functional information. Since the size and shape of structures vary across modalities, it is more challenging to evaluate the alignment quality. This has led to the use of similarity measures such as mutual information.
- Characterizing a population of subjects. In contrast to intra-subject registration, a one-to-one mapping may not exist between subjects, depending on the structural variability of the organ of interest. Inter-subject registration is required for atlas construction in computational anatomy. Here, the objective is to statistically model the anatomy of organs across subjects.

There are several important considerations when performing image registration:

- The transformation model. Common choices are rigid, affine, and deformable transformation models. B-spline and thin plate spline models are commonly used for parameterized transformation fields. Non-parametric or dense deformation fields carry a displacement vector at every grid location; this necessitates additional regularization constraints. A specific class of deformation fields are diffeomorphisms, which are invertible transformations with a smooth inverse.
- The similarity metric. A distance or similarity function is used to quantify the registration quality. This similarity can be calculated either on the original images or on features extracted from the images. Common similarity measures are sum of squared distances (SSD), correlation coefficient, and mutual information. The choice of similarity measure depends on whether the images are from the same modality; the acquisition noise can also play a role in this decision. For example, SSD is the optimal similarity measure for images of the same modality with Gaussian noise. However, the image statistics in ultrasound are significantly different from Gaussian noise, leading to the introduction of ultrasound specific similarity measures. Multi-modal registration requires a more sophisticated similarity measure; alternatively, a different image representation can be used, such as structural representations.
- The optimization procedure. Either continuous or discrete optimization is performed. For continuous optimization, gradient-based optimization techniques are applied to improve the convergence speed.

Visualization

Visualization plays several key roles in Medical Image Computing. Methods from scientific visualization are used to understand and communicate about medical images, which are inherently spatial-temporal. Data visualization and data analysis are used on unstructured data forms, for example when evaluating statistical measures derived during algorithmic processing. Direct interaction with data, a key feature of the visualization process, is used to perform visual queries about data, annotate images, guide segmentation and registration processes, and control the visual representation of data (by controlling lighting rendering properties and viewing parameters). Visualization is used both for initial exploration and for conveying intermediate and final results of analyses.

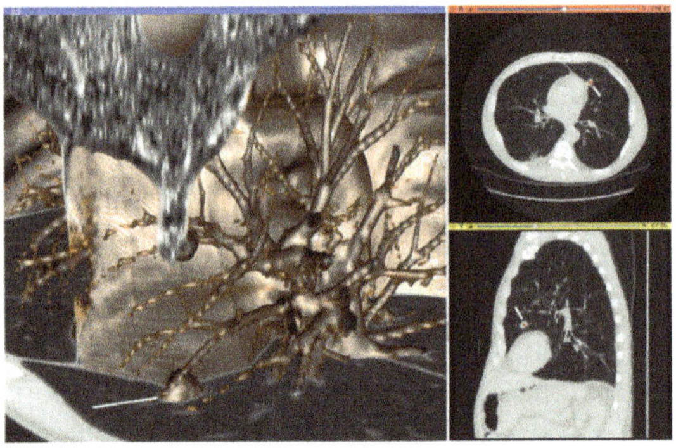

Volume rendering (left), axial cross-section (right top), and sagittal cross-section (right bottom) of a CT image of a subject with multiple nodular lesions (white line) in the lung.

The figure "Visualization of Medical Imaging" illustrates several types of visualization: 1. the display of cross-sections as gray scale images; 2. reformatted views of gray scale images (the sagittal view in this example has a different orientation than the original direction of the image acquisition; and 3. A 3D volume rendering of the same data. The nodular lesion is clearly visible in the different presentations and has been annotated with a white line.

Atlases

Medical images can vary significantly across individuals due to people having organs of different shapes and sizes. Therefore, representing medical images to account for this variability is crucial. A popular approach to represent medical images is through the use of one or more atlases. Here, an atlas refers to a specific model for a population of images with parameters that are learned from a training dataset.

The simplest example of an atlas is a mean intensity image, commonly referred to as a template. However, an atlas can also include richer information, such as local image statistics and the probability that a particular spatial location has a certain label. New medical images, which are not used during training, can be mapped to an atlas, which has been tailored to the specific application, such as segmentation and group analysis. Mapping an image to an atlas usually involves registering the image and the atlas. This deformation can be used to address variability in medical images.

Single Template

The simplest approach is to model medical images as deformed versions of a single template image. For example, anatomical MRI brain scans are often mapped to the MNI template as to represent all the brain scans in common coordinates. The main drawback of a single-template approach is that if there are significant differences between the template and a given test image, then there may not be a good way to map one onto the other. For example, an anatomical MRI brain scan of a patient with severe brain abnormalities (i.e., a tumor or surgical procedure), may not easily map to the MNI template.

Multiple Templates

Rather than relying on a single template, multiple templates can be used. The idea is to represent an image as a deformed version of one of the templates. For example, there could be one template for a healthy population and one template for a diseased population. However, in many applications, it is not clear how many templates are needed. A simple albeit computationally expensive way to deal with this is to have every image in a training dataset be a template image and thus every new image encountered is compared against every image in the training dataset. A more recent approach automatically finds the number of templates needed.

Statistical Analysis

Statistical methods combine the medical imaging field with modern Computer Vision, Machine Learning and Pattern Recognition. Over the last decade, several large datasets have been made publicly available (see for example ADNI, 1000 functional Connectomes Project), in part due to collaboration between various institutes and research centers. This increase in data size calls for new algorithms that can mine and detect subtle changes in the images to address clinical questions. Such clinical questions are very diverse and include group analysis, imaging biomarkers, disease phenotyping and longitudinal studies.

Group Analysis

In the Group Analysis, the objective is to detect and quantize abnormalities induced by a disease by comparing the images of two or more cohorts. Usually one of these cohorts consist of normal (control) subjects, and the other one consists of abnormal patients. Variation caused by the disease can manifest itself as abnormal deformation of anatomy (see Voxel-based morphometry). For example, shrinkage of sub-cortical tissues such as the Hippocampus in brain may be linked to Alzheimer's disease. Additionally, changes in biochemical (functional) activity can be observed using imaging modalities such as Positron Emission Tomography.

The comparison between groups is usually conducted on the voxel level. Hence, the most popular pre-processing pipeline, particularly in neuroimaging, transforms all of the images in a dataset to a common coordinate frame via (Medical Image Registration) in order to maintain correspondence between voxels. Given this voxel-wise correspondence, the most common Frequentist method is to extract a statistic for each voxel (for example, the mean voxel intensity for each group) and perform statistical hypothesis testing to evaluate whether a null hypothesis is or is not supported. The null hypothesis typically assumes that the two cohorts are drawn from the same distribution, and hence, should have the same statistical properties (for example, the mean values of two groups are equal for a particular voxel). Since medical images contain large numbers of voxels, the issue of multiple comparison needs to be addressed,. There are also Bayesian approaches to tackle group analysis problem.

Classification

Although group analysis can quantify the general effects of a pathology on an anatomy and function, it does not provide subject level measures, and hence cannot be used as biomarkers for diagnosis (see Imaging Biomarkers). Clinicians, on the other hand, are often interested in early diagnosis of the pathology (i.e. classification,) and in learning the progression of a disease (i.e. regression). From methodological point of view, current techniques varies from applying standard machine learning algorithms to medical imaging datasets (e.g. Support Vector Machine), to developing new approaches adapted for the needs of the field. The main difficulties are as follows:

- Small sample size (Curse of Dimensionality): a large medical imaging dataset contains hundreds to thousands of images, whereas the number of voxels in a typical volumetric image can easily go beyond millions. A remedy to this problem is to reduce the number of features in an informative sense (see dimensionality reduction). Several unsupervised and semi-/supervised, approaches have been proposed to address this issue.
- Interpretability: A good generalization accuracy is not always the primary objective, as clinicians would like to understand which parts of anatomy are affected by the disease. Therefore, interpretability of the results is very important; methods that ignore the image structure are not favored. Alternative methods based on feature selection have been proposed,.

Clustering

Image-based pattern classification methods typically assume that the neurological effects of a disease are distinct and well defined. This may not always be the case. For a number of medical conditions, the patient populations are highly heterogeneous, and further categorization into sub-conditions has not been established. Additionally, some diseases (e.g., Autism Spectrum Disorder (ASD), Schizophrenia, Mild cognitive impairment (MCI)) can be characterized by a continuous or nearly-continuous spectra from mild cognitive impairment to very pronounced pathological changes. To facilitate image-based analysis of heterogeneous disorders, methodological alternatives to pattern classification have been developed. These techniques borrow ideas from high-dimensional clustering and high-dimensional pattern-regression to cluster a given population into homogeneous sub-populations. The goal is to provide a better quantitative understanding of the disease within each sub-population.

Shape Analysis

Shape Analysis is the field of Medical Image Computing that studies geometrical properties of structures obtained from different imaging modalities. Shape analysis recently become of increasing interest to the medical community due to its potential to precisely locate morphological changes between different populations of structures, i.e. healthy vs pathological, female vs male, young vs elderly. Shape Analysis includes two main steps: shape correspondence and statistical analysis.

- Shape correspondence is the methodology that computes correspondent locations between geometric shapes represented by triangle meshes, contours, point sets or volumetric images. Obviously definition of correspondence will influence directly the analysis. Among the different options for correspondence frameworks we can find: Anatomical correspondence, manual landmarks, functional correspondence (i.e. in brain morphometry locus responsible for same neuronal functionality), geometry correspondence, (for image volumes) intensity similarity, etc. Some approaches, e.g. spectral shape analysis, do not require correspondence but compare shape descriptors directly.
- Statistical analysis will provide measurements of structural change at correspondent locations.

Longitudinal Studies

In longitudinal studies the same person is imaged repeatedly. This information can be incorporated both into the image analysis, as well as into the statistical modeling.

- In longitudinal image processing, segmentation and analysis methods of individual time points are informed and regularized with common information usually from a within-subject template. This regularization is designed to reduce measurement noise and thus helps increase sensitivity and statistical power. At the same time over-regularization needs to be avoided, so that effect sizes remain stable. Intense regularization, for example, can lead to excellent test-retest reliability, but limits the ability to detect any true changes and differences across groups. Often a trade-off needs to be aimed for, that optimizes noise reduction at the cost of limited effect size loss. Another common challenge in longitudinal image processing is the, often unintentional, introduction of processing bias. When, for example, follow-up images get registered and resampled to the baseline image, interpolation artifacts get introduced to only the follow-up images and not the baseline. These artifact can cause spurious effects (usually a bias towards overestimating longitudinal change and thus underestimating required sample size). It is therefore essential that all-time points get treated exactly the same to avoid any processing bias.
- Post-processing and statistical analysis of longitudinal data usually requires dedicated statistical tools such as repeated measure ANOVA or the more powerful linear mixed effects models. Additionally, it is advantageous to consider the spatial distribution of the signal. For example, cortical thickness measurements will show a correlation within-subject across time and also within a neighborhood on the cortical surface - a fact that can be used to increase statistical power. Furthermore, time-to-even (aka survival) analysis is frequently employed to analyze longitudinal data and determine significant predictors.

Image-Based Physiological Modelling

Traditionally, medical image computing has seen to address the quantification and fusion of structural or functional information available at the point and time of image acquisition. In this regard, it can be seen as quantitative sensing of the underlying anatomical, physical or physiological processes. However, over the last few years, there has been a growing interest in the predictive assessment of disease or therapy course. Image-based modelling, be it of biomechanical or physiological nature, can therefore extend the possibilities of image computing from a descriptive to a predictive angle. According to the STEP research roadmap, the Virtual Physiological Human (VPH) is a methodological and technological framework that, once established, will enable the investigation of the human body as a single complex system. Underlying the VPH concept, the International Union for Physiological Sciences (IUPS) has been sponsoring the IUPS Physiome Project for more than a decade,. This is a worldwide public domain effort to provide a computational framework for understanding human physiology. It aims at developing integrative models at all levels of biological organization, from genes to the whole organisms via gene regulatory networks, protein pathways, integrative cell functions, and tissue and whole organ structure/function relations. Such an approach aims at transforming current practice in medicine and underpins a new era of computational medicine.

In this context, medical imaging and image computing play, and will continue to play, an increasingly important role as they provide systems and methods to image, quantify and fuse both structural and functional information about the human being in vivo. These two broad research areas include the transformation of generic computational models to represent specific subjects, thus paving the way for personalized computational models. Individualization of generic computational models through imaging can be realized in three complementary directions:

- definition of the subject-specific computational domain (anatomy) and related subdomains (tissue types);
- definition of boundary and initial conditions from (dynamic and/or functional) imaging; and
- characterization of structural and functional tissue properties.

In addition, imaging also plays a pivotal role in the evaluation and validation of such models both in humans and in animal models, and in the translation of models to the clinical setting with both diagnostic and therapeutic applications. In this specific context, molecular, biological, and pre-clinical imaging render additional data and understanding of basic structure and function in molecules, cells, tissues and animal models that may be transferred to human physiology where appropriate.

The applications of image-based VPH/Physiome models in basic and clinical domains are vast. Broadly speaking, they promise to become new *virtual imaging techniques*. Effectively more, often non-observable, parameters will be imaged *in silico* based on the integration of observable but sometimes sparse and inconsistent multimodal images and physiological measurements. Computational models will serve to engender interpretation of the measurements in a way compliant with the underlying biophysical, biochemical or biological laws of the physiological or pathophysiological processes under investigation. Ultimately, such investigative tools and systems will help our understanding of disease processes, the natural history of disease evolution, and the influence on the course of a disease of pharmacological and/or interventional therapeutic procedures. Cross-fertilization between imaging and modelling goes beyond interpretation of measurements in a way consistent with physiology. Image-based patient-specific modelling, combined with models of medical devices and pharmacological therapies, opens the way to predictive imaging whereby one will be able to understand, plan and optimize such interventions *in silico*.

Mathematical Methods in Medical Imaging

A number of sophisticated mathematical methods have entered medical imaging, and have already been implemented in various software packages. These include approaches based on partial differential equations (PDEs) and curvature driven flows for enhancement, segmentation, and registration. Since they employ PDEs, the methods are amenable to parallelization and implementation on GPGPUs. A number of these techniques have been inspired from ideas in optimal control. Accordingly, very recently ideas from control have recently made their way into interactive methods, especially segmentation. Moreover, because of noise and the need for statistical estimation techniques for more dynamically changing imagery, the Kalman filter and particle filter have come into use. A survey of these methods with an extensive list of references may be found in.

Modality Specific Computing

Some imaging modalities provide very specialized information. The resulting images cannot be treated as regular scalar images and give rise to new sub-areas of Medical Image Computing. Examples include diffusion MRI, functional MRI and others.

Diffusion MRI

Diffusion MRI is a structural magnetic resonance imaging modality that allows measurement of the diffusion process of molecules. Diffusion is measured by applying a gradient pulse to a magnetic field along a particular direction. In a typical acquisition, a set of uniformly distributed gradient directions is used to create a set of diffusion weighted volumes. In addition, an unweighted volume is acquired under the same magnetic field without application of a gradient pulse. As each acquisition is associated with multiple volumes, diffusion MRI has created a variety of unique challenges in medical image computing.

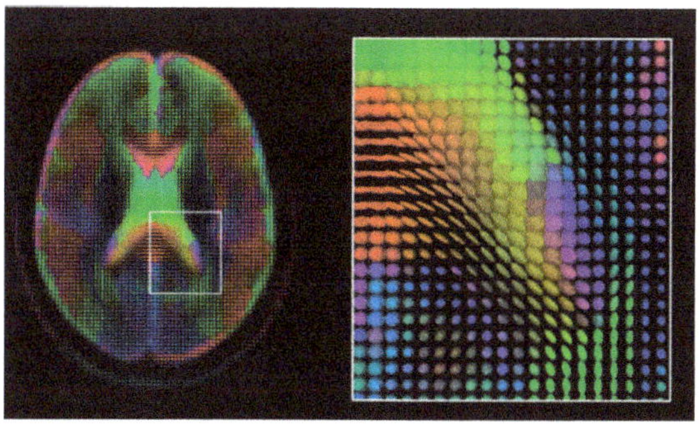

A mid-axial slice of the ICBM diffusion tensor image template. Each voxel's value is a tensor represented here by an ellipsoid. Color denotes principal orientation: red = left-right, blue=inferior-superior, green = posterior-anterior

Diffusion MRI is a structural magnetic resonance imaging modality that allows measurement of the diffusion process of molecules. Diffusion is measured by applying a gradient pulse to a magnetic field along a particular direction. In a typical acquisition, a set of uniformly distributed gradient directions is used to create a set of diffusion weighted volumes. In addition, an unweighted volume is acquired under the same magnetic field without application of a gradient pulse. As each acquisition is associated with multiple volumes, diffusion MRI has created a variety of unique challenges in medical image computing.

In medicine, there are two major computational goals in diffusion MRI:

- Estimation of local tissue properties, such as diffusivity;
- Estimation of local directions and global pathways of diffusion.

The diffusion tensor, a 3×3 symmetric positive-definite matrix, offers a straightforward solution to both of these goals. It is proportional to the covariance matrix of a Normally distributed local diffusion profile and, thus, the dominant eigenvector of this matrix is the principal direction of local diffusion. Due to the simplicity of this model, a maximum likelihood estimate of the diffusion tensor can be found by simply solving a system of linear equations at each location independently. However, as the volume is assumed to contain contiguous tissue fibers, it may be preferable to estimate the volume of diffusion tensors in its entirety by imposing regularity conditions on the underlying field of tensors. Scalar values can be extracted from the diffusion tensor, such as the fractional anisotropy, mean, axial and radial diffusivities, which *indirectly* measure tissue properties such as the dysmyelination of axonal fibers or the presence of edema.Standard scalar image computing methods, such as registration and segmentation, can be applied directly to volumes of such scalar values. However, to fully exploit the information in the diffusion tensor, these methods have been adapted to account for tensor valued volumes when performing registration and segmentation.

Given the principal direction of diffusion at each location in the volume, it is possible to estimate the global pathways of diffusion through a process known as tractography. However, due to the relatively low resolution of diffusion MRI, many of these pathways may cross, kiss or fan at a single location. In this situation, the single principal direction of the diffusion tensor is not an appropriate model for the local diffusion distribution. The most common solution to this problem is to estimate multiple directions of local diffusion using more complex models. These include mixtures of diffusion tensors, Q-ball imaging, diffusion spectrum imaging and fiber orientation distribution functions, which typically require HARDI acquisition with a large number of gradient directions. As with the diffusion tensor, volumes valued with these complex models require special treatment when applying image computing methods, such as registration and segmentation.

Functional MRI

Functional magnetic resonance imaging (fMRI) is a medical imaging modality that indirectly measures neural activity by observing the local hemodynamics, or blood oxygen level dependent signal (BOLD). fMRI data offers a range of insights, and can be roughly divided into two categories:

- **Task related fMRI** is acquired as the subject is performing a sequence of timed experimental conditions. In block-design experiments, the conditions are present for short periods of time (e.g., 10 seconds) and are alternated with periods of rest. Event-related experiments rely on a random sequence of stimuli and use a single time point to denote each condition. The standard approach to analyze task related fMRI is the general linear model (GLM)
- **Resting state fMRI** is acquired in the absence of any experimental task. Typically, the objective is to study the intrinsic network structure of the brain. Observations made during rest have also been linked to specific cognitive processes such as encoding or reflection. Most studies of resting state fMRI focus on low frequency fluctuations of the fMRI signal (LF-BOLD). Seminal discoveries include the default network, a comprehensive cortical parcellation, and the linking of network characteristics to behavioral parameters.

There is a rich set of methodology used to analyze functional neuroimaging data, and there is often no consensus regarding the *best* method. Instead, researchers approach each problem independently and select a suitable model/algorithm. In this context there is a relatively active exchange among neuroscience, computational biology, statistics, and machine learning communities. Prominent approaches include

- **Massive univariate approaches** that probe individual voxels in the imaging data for a relationship to the experiment condition. The prime approach is the general linear model (GLM)
- **Multivariate- and classifier based approaches**, often referred to as multi voxel pattern analysis or multi-variate pattern analysis probe the data for global and potentially distributed responses to an experimental condition. Early approaches used support vector machines (SVM) to study responses to visual stimuli. Recently, alternative pattern recognition algorithms have been explored, such as random forest based gini contrast or sparse regression and dictionary learning
- **Functional connectivity analysis** studies the intrinsic network structure of the brain, including the interactions between regions. The majority of such studies focus on resting state data to parcelate the brain or to find correlates to behavioral measures. Task specific data can be used to study causal relationships among brain regions (e.g., dynamic causal mapping (DCM)).

When working with large cohorts of subjects, the normalization (registration) of individual subjects into a common reference frame is crucial. A body of work and tools exist to perform normalization based on anatomy (FSL, FreeSurfer, SPM). Alignment taking spatial variability across subjects into account is a more recent line of work. Examples are the alignment of the cortex based on fMRI signal correlation, the alignment based on the global functional connectivity structure both in task-, or resting state data, and the alignment based on stimulus specific activation profiles of individual voxels.

Software

Software for medical image computing is a complex combination of systems providing

IO, visualization and interaction, user interface, data management and computation. Typically system architectures are layered to serve algorithm developers, application developers, and users. The bottom layers are often libraries and/or toolkits which provide base computational capabilities; while the top layers are specialized applications which address specific medical problems, diseases, or body systems.

Additional Notes

Medical Image Computing is also related to the field of Computer Vision. An international society, the MICCAI society represents the field and organizes an annual conference and associated workshops. Proceedings for this conference are published by Springer in the Lecture Notes in Computer Science series. In 2000, N. Ayache and J. Duncan reviewed the state of the field.

MIND UPLOADING

Whole brain emulation (WBE) or mind uploading (sometimes called "mind copying" or "mind transfer") is the hypothetical process of scanning mental state (including long-term memory and "self") of a particular brain substrate and copying it to a computational device, such as a digital, analog, quantum-based or software-based artificial neural network. The computational device could then run a simulation model of the brain's information processing, such that it responds in essentially the same way as the original brain (i.e., indistinguishable from the brain for all relevant purposes) and experiences having a conscious mind.

Mind uploading may potentially be accomplished by either of two methods: Copy-and-Transfer or gradual replacement of neurons. In the case of the former method, mind uploading would be achieved by scanning and mapping the salient features of a biological brain, and then by copying, transferring, and storing that information state into a computer system or another computational device. The simulated mind could be within a virtual reality or simulated world, supported by an anatomic 3D body simulation model. Alternatively, the simulated mind could reside in a computer that is inside (or connected to) a (not necessarily humanoid) robot or a biological body in real life.

Among some futurists and within the transhumanist movement, mind uploading is treated as an important proposed life extension technology. Some believe mind uploading is humanity's current best option for preserving the identity of the species, as opposed to cryonics. Another aim of mind uploading is to provide a permanent backup to our "mind-file", and a means for functional copies of human minds to survive a global disaster or interstellar space travels. Whole brain emulation is discussed by some futurists as a "logical endpoint" of the topical computational neuroscience and neuroinformatics fields, both about brain simulation for medical research purposes. It is discussed in artificial intelligence research publications as an approach to strong AI. Computer-based intelligence such as an upload could think much faster than a biological human even if it were no more intelligent. A large-scale society of uploads might, according to futurists, give rise to a technological singularity, meaning a sudden time constant decrease in the exponential development of technology. Mind uploading is a central conceptual feature of numerous science fiction novels and films. Substantial mainstream research in related areas is being conducted in animal brain mapping and simulation, development of faster super computers, virtual reality, brain–computer interfaces, connectomics and information extraction from dynamically functioning brains. According to supporters, many of the tools and ideas needed to achieve mind uploading already exist or are currently under active development; however, they will admit that others are, as yet, very speculative, but still in the realm of engineering possibility. Neuroscientist Randal Koene has formed a nonprofit organization called Carbon Copies to promote mind uploading research.

Overview

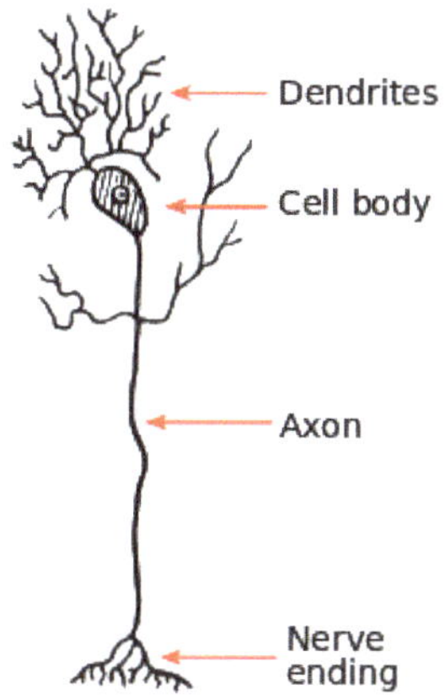

Neuron anatomical model

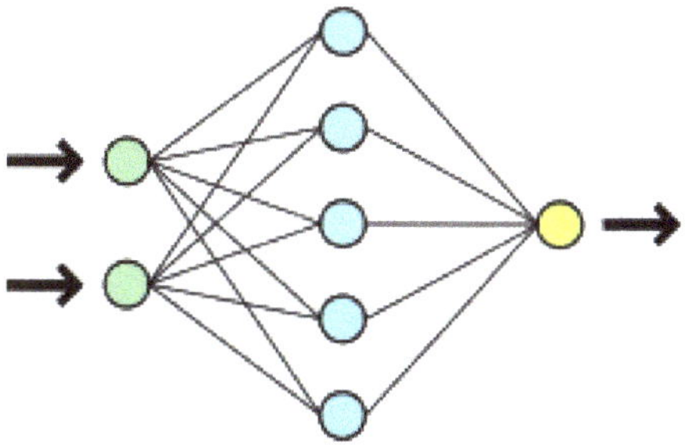

Simple artificial neural network

The human brain contains about 86 billion nerve cells called neurons, each individually linked to other neurons by way of connectors called axons and dendrites. Signals at the junctures (synapses) of these connections are transmitted by the release and detection of chemicals known as neurotransmitters. The established neuroscientific consensus is that the human mind is largely an emergent property of the information processing of this neural network.

Importantly, neuroscientists have stated that important functions performed by the mind, such as learning, memory, and consciousness, are due to purely physical and electrochemical processes in the brain and are governed by applicable laws. For example, Christof Koch and Giulio Tononi wrote in *IEEE Spectrum*:

"Consciousness is part of the natural world. It depends, we believe, only on mathematics and logic and on the imperfectly known laws of physics, chemistry, and biology; it does not arise from some magical or otherworldly quality."

The concept of mind uploading is based on this mechanistic view of the mind, and denies the vitalist view of human life and consciousness.

Eminent computer scientists and neuroscientists have predicted that specially programmed computers will be capable of thought and even attain consciousness, including Koch and Tononi, Douglas Hofstadter, Jeff Hawkins, Marvin Minsky, Randal A. Koene, and Rodolfo Llinas.

Such an artificial intelligence capability might provide a computational substrate necessary for uploading.

However, even though uploading is dependent upon such a general capability, it is conceptually distinct from general forms of AI in that it results from dynamic reanimation of information derived from a specific human mind so that the mind retains a sense of historical identity (other forms are possible but would compromise or eliminate the life-extension feature generally associated with uploading). The transferred and reanimated information would become a form of artificial intelligence, sometimes called an infomorph or *"noömorph"*.

Many theorists have presented models of the brain and have established a range of estimates of the amount of computing power needed for partial and complete simulations. Using these models, some have estimated that uploading may become possible within decades if trends such as Moore's law continue.

Theoretical Benefits and Applications

"Immortality" or Backup

In theory, if the information and processes of the mind can be disassociated from the biological body, they are no longer tied to the individual limits and lifespan of that body. Furthermore, information within a brain could be partly or wholly copied or transferred to one or more other substrates (including digital storage or another brain), thereby – from a purely mechanistic perspective – reducing or eliminating "mortality risk" of such information. This general proposal appears to have been first made in the biomedical literature in 1971 by biogerontologistGeorge M. Martin of the University of Washington.

Uploaded Astronaut

An "uploaded astronaut" is the application of mind uploading to human spaceflight. An uploaded astronaut would consist of a human mental content transferred or copied to a space humanoid robot or a spacecraft's data storage device. This would eliminate the harms caused by a zero gravity environment, the vacuum of space and cosmic radiation to the human body since both a humanoid robot and a spacecraft can be more resistant than a biological entity in such conditions, permitting longer and farther voyages through outer space than manned spaceflight. Furthermore, an uploaded astronaut may not require a large spacecraft so spacecrafts at the scale of the StarChip might suffice. Alien uploaded astronauts are conceivable as well. In science fiction, Charlie Stross's *Accelerando* features a can-sized starship that visits a nearby star system with an "e-crew" of 63 uploaded astronauts.

Relevant Technologies and Techniques

The focus of mind uploading, in the case of copy-and-transfer, is on data acquisition, rather than data maintenance of the brain. A set of approaches known as loosely coupled off-loading (LCOL) may be used in the attempt to characterize and copy the mental contents of a brain. The LCOL approach may take advantage of self-reports, life-logs and video recordings that can be analyzed by artificial intelligence. A bottom-up approach may focus on the specific resolution and morphology of neurons, the spike times of neurons, the times at which neurons produce action potential responses.

Computational Complexity

Advocates of mind uploading point to Moore's law to support the notion that the necessary computing power is expected to become available within a few decades. However, the actual computational requirements for running an uploaded human mind are very difficult to quantify, potentially rendering such an argument specious.

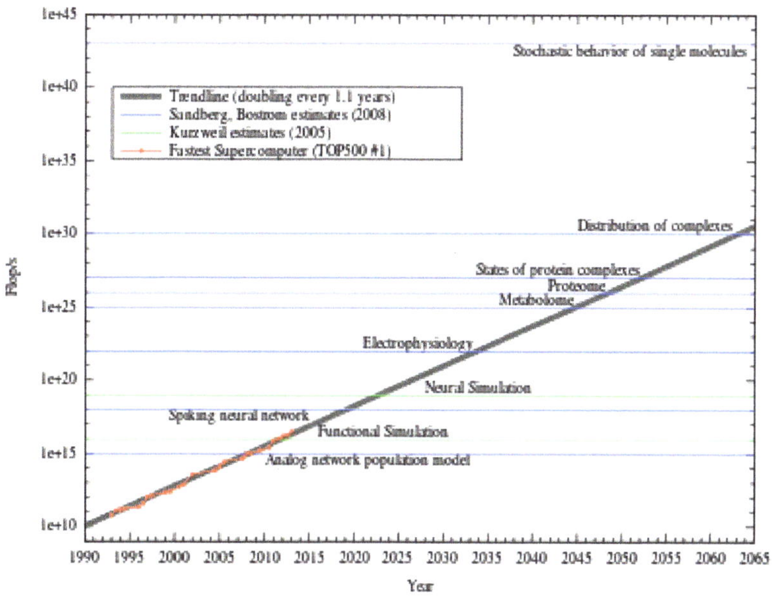

Estimates of how much processing power is needed to emulate a human brain at various levels (from Ray Kurzweil and the chart to the left), along with the fastest supercomputer from TOP500 mapped by year. Note the logarithmic scale and exponential trendline, which assumes the computational capacity doubles every 1.1 years. Kurzweil believes that mind uploading will be possible at neural simulation, while the Sandberg, Bostrom report is less certain about where consciousness arises.

Regardless of the techniques used to capture or recreate the function of a human mind, the processing demands are likely to be immense, due to the large number of neurons in the human brain along with the considerable complexity of each neuron.

In 2004, Henry Markram, lead researcher of the "Blue Brain Project", stated that "it is not their goal to build an intelligent neural network", based solely on the computational demands such a project would have.

It will be very difficult because, in the brain, every molecule is a powerful computer and we would need to simulate the structure and function of trillions upon trillions of molecules as well as all the rules that govern how they interact. You would literally need computers that are trillions of times bigger and faster than anything existing today.

Five years later, after successful simulation of part of a rat brain, the same scientist was much more bold and optimistic. In 2009, when he was director of the Blue Brain Project, he claimed that

A detailed, functional artificial human brain can be built within the next 10 years

Required computational capacity strongly depend on the chosen level of simulation model scale:

Level	CPU demand (FLOPS)	Memory demand (Tb)	$1 million super-computer (Earliest year of making)
Analog network population model	10^{15}	10^{2}	2008
Spiking neural network	10^{18}	10^{4}	2019
Electrophysiology	10^{22}	10^{4}	2033
Metabolome	10^{25}	10^{6}	2044
Proteome	10^{26}	10^{7}	2048
States of protein complexes	10^{27}	10^{8}	2052
Distribution of complexes	10^{30}	10^{9}	2063
Stochastic behavior of single molecules	10^{43}	10^{14}	2111

Simulation Model Scale

Since the function of the human mind, and how it might arise from the working of the brain's neural network, are poorly understood issues, mind uploading relies on the idea of neural network emulation. Rather than having to understand the high-level psychological processes and large-scale structures of the brain, and model them using classical artificial intelligence methods and cognitive psychology models, the low-level structure of the underlying neural network is captured, mapped and emulated with a computer system.

In computer science terminology, rather than analyzing and reverse engineering the behavior of the algorithms and data structures that resides in the brain, a blueprint of its source code is translated to another programming language. The human mind and the personal identity then, theoretically, is generated by the emulated neural network in an identical fashion to it being generated by the biological neural network.

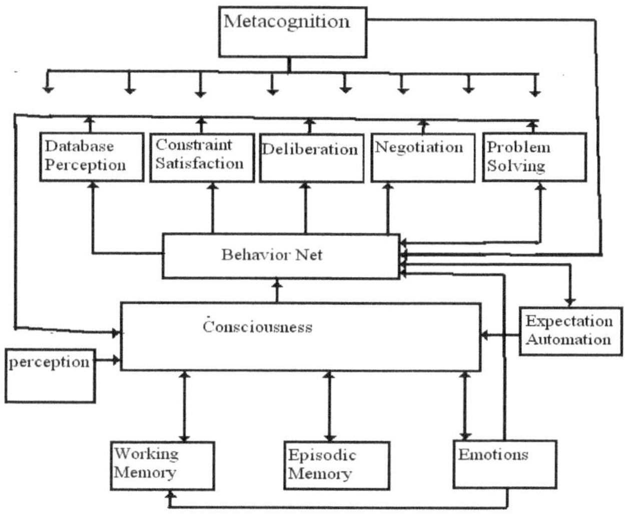

Franklin IDA Architecture

A high-level cognitive AI model of the brain architecture is not required for brain emulation

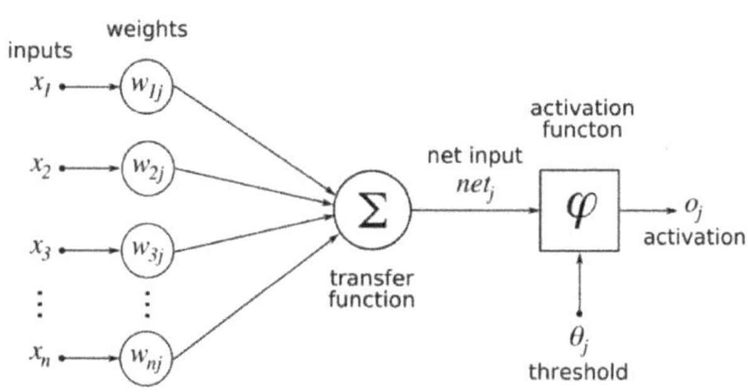

Simple neuron model: Black-box dynamic non-linear signal processing system

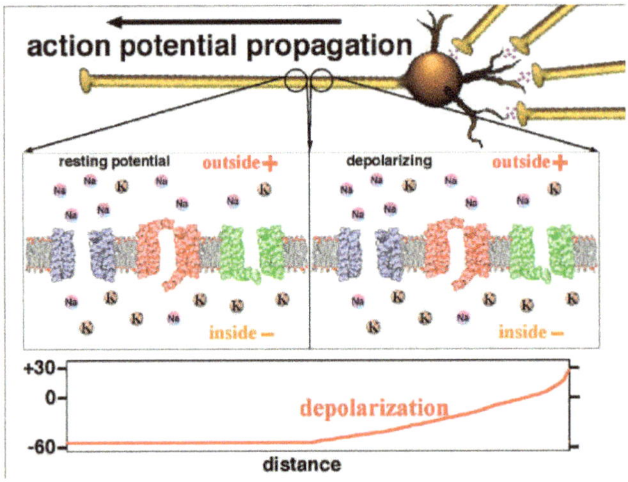

Metabolism model: The movement of positively charged ions through the ion channels controls the membrane electrical action potential in an axon.

On the other hand, a molecule-scale simulation of the brain is not expected to be required, provided that the functioning of the neurons is not affected by quantum mechanical processes. The neural network emulation approach only requires that the functioning and interaction of neurons and synapses are understood. It is expected that it is sufficient with a black-box signal processing model of how the neurons respond to nerve impulses (electrical as well as chemical synaptic transmission).

A sufficiently complex and accurate model of the neurons is required. A traditional artificial neural network model, for example multi-layer perceptron network model, is not considered as sufficient. A dynamic spiking neural network model is required, which reflects that the neuron fires only when a membrane potential reaches a certain level. It is likely that the model must include delays, non-linear functions and differential equations describing the relation between electrophysical parameters such as electrical currents, voltages, membrane states (ion channel states) and neuromodulators.

Since learning and long-term memory are believed to result from strengthening or weakening the synapses via a mechanism known as synaptic plasticity or synaptic adaptation, the model should include this mechanism. The response of sensory receptors to various stimuli must also be modelled.

Furthermore, the model may have to include metabolism, i.e. how the neurons are affected by hormones and other chemical substances that may cross the blood–brain barrier. It is considered likely that the model must include currently unknown neuromodulators, neurotransmitters and ion channels. It is considered unlikely that the simulation model has to include protein interaction, which would make it computationally complex.

A digital computer simulation model of an analog system such as the brain is an approximation that introduces random quantization errors and distortion. However, the biological neurons also suffer from randomness and limited precision, for example due to background noise. The errors of the discrete model can be made smaller than the randomness of the biological brain by choosing a sufficiently high variable resolution and sample rate, and sufficiently accurate models of non-linearities. The computational power and computer memory must however be sufficient to run such large simulations, preferably in real time.

Scanning and Mapping Scale of an Individual

When modelling and simulating the brain of a specific individual, a brain map or connectivity database showing the connections between the neurons must be extracted from an anatomic model of the brain. For whole brain simulation, this network map should show the connectivity of the whole nervous system, including the spinal cord, sensory receptors, and muscle cells. Destructive scanning of a small sample of tissue from a mouse brain including synaptic details is possible as of 2010.

However, if short-term memory and working memory include prolonged or repeated firing of neurons, as well as intra-neural dynamic processes, the electrical and chemical signal state of the synapses and neurons may be hard to extract. The uploaded mind may then perceive a memory loss of the events and mental processes immediately before the time of brain scanning. A full brain map has been estimated to occupy less than 2×10^{16} bytes (20,000 TB) and would store the addresses of the connected neurons, the synapse type and the synapse "weight" for each of the brains' 10^{15} synapses. However, the biological complexities of true brain function (e.g. the epigenetic states of neurons, protein components with multiple functional states, etc.) may preclude an accurate prediction of the volume of binary data required to faithfully represent a functioning human mind.

Serial Sectioning

A possible method for mind uploading is serial sectioning, in which the brain tissue and perhaps other parts of the nervous system are frozen and then scanned and analyzed layer by layer, which for frozen samples at nano-scale requires a cryo-ultramicrotome, thus capturing the structure of the neurons and their interconnections. The exposed surface of frozen nerve tissue would be scanned and recorded, and then the surface layer of tissue removed. While this would be a very slow and labor-intensive process, research is currently underway to automate the collection and microscopy of serial sections. The scans would then be analyzed, and a model of the neural net recreated in the system that the mind was being uploaded into. There are uncertainties with this approach using current microscopy techniques. If it is possible to replicate neuron function from its visible structure alone, then the resolution afforded by a scanning electron microscope would suffice for such a technique. However, as the function of brain tissue is partially determined by molecular events (particularly at synapses, but also at other places on the neuron's cell membrane), this may not suffice for capturing and simulating neuron functions. It may be possible to extend the techniques of serial sectioning and to capture the internal molecular makeup of neurons, through the use of sophisticated immunohistochemistry staining methods that could then be read via confocal laser scanning microscopy. However, as the physiological genesis of 'mind' is not currently known, this method may not be able to access all of the necessary biochemical information to recreate a human brain with sufficient fidelity.

Brain Imaging

It may also be possible to create functional 3D maps of the brain activity, using advanced neuroimaging technology, such as functional MRI (fMRI, for mapping change in blood flow), magnetoencephalography (MEG, for mapping of electrical currents), or combinations of multiple methods, to build a detailed three-dimensional model of the brain using non-invasive and non-destructive methods. Today, fMRI is often combined with MEG for creating functional maps of human cortex during more complex cognitive tasks, as the methods complement each other. Even though current imaging technology lacks the spatial resolution needed to gather the information needed for such a scan, important recent and future developments are predicted to substantially improve both spatial and temporal resolutions of existing technologies.

Brain Simulation

There is ongoing work in the field of brain simulation, including partial and whole simulations of some animals. For example, the *C. elegans* roundworm, *Drosophila* fruit fly, and mouse have all been simulated to various degrees. The Blue Brain Project by the Brain and Mind Institute of the *École Polytechnique Fédérale de Lausanne,* Switzerland is an attempt to create a synthetic brain by reverse-engineering mammalian brain circuitry.

ISSUES

Philosophical Issues

Underlying the concept of "mind uploading" (more accurately "mind transferring") is the broad philosophy that consciousness lies within the brain's information processing and is in essence an emergent feature that arises from large neural network high-level patterns of organization, and that the same patterns of organization can be realized in other processing devices. Mind uploading also relies on the idea that the human mind (the "self" and the long-term memory), just like non-human minds, is represented by the current neural network paths and the weights of the brain synapses rather than by a dualistic and mystic soul and spirit. The mind or "soul" can be defined as the information state of the brain, and is immaterial only in the same sense as the information content of a data file or the state of a computer software currently residing in the work-space memory of the computer. Data specifying the information state of the neural network can be captured and copied as a "computer file" from the brain and re-implemented into a different physical form. This is not to deny that minds are richly adapted to their substrates. An analogy to the idea of mind uploading is to copy the temporary information state (the variable values) of a computer program from the computer memory to another computer and continue its execution. The other computer may perhaps have different hardware architecture but emulates the hardware of the first computer.

These issues have a long history. In 1775 Thomas Reid wrote:

I would be glad to know... whether when my brain has lost its original structure, and when some hundred years after the same materials are fabricated so curiously as to become an intelligent being, whether, I say that being will be me; or, if, two or three such beings should be formed out of my brain; whether they will all be me, and consequently one and the same intelligent being.

A considerable portion of transhumanists and singularitarians place great hope into the belief that they may become immortal, by creating one or many non-biological functional copies of their brains, thereby leaving their "biological shell". However, the philosopher and transhumanist Susan Schneider claims that at best, uploading would create a copy of the original persons mind. Susan Schneider agrees that consciousness has a computational basis, but this doesn't mean we can upload and survive. According to her views, "uploading" would probably result in the death of the original person's brain, while only outside observers can maintain the illusion of the original person still being alive. For it is implausible to think that one's consciousness would leave one's brain and travel to a remote location; ordinary physical objects do not behave this way. Ordinary objects (rocks, tables, etc.) are not simultaneously here, and somewhere else. At best, a copy of the original mind is created. Others have argued against such conclusions. For example, Buddhist transhumanist James Hughes has pointed out that this consideration only goes so far: if one believes the self is an illusion, worries about survival are not reasons to avoid uploading, and Keith Wiley has presented an argument wherein all resulting minds of an uploading procedure are granted equal primacy in their claim to the original identity, such that survival of the self is determined retroactively from a strictly subjective position.

Another potential consequence of mind uploading is that the decision to "upload" may then create a mindless symbol manipulator instead of a conscious mind (see philosophical zombie). Are we to assume that an upload is conscious if it displays behaviors that are highly indicative of consciousness? Are we to assume that an upload is conscious if it verbally insists that it is conscious? Could there be an absolute upper limit in processing speed above which consciousness cannot be sustained? The mystery of consciousness precludes a definitive answer to this question. Numerous scientists, including Kurzweil, strongly believe that determining whether a separate entity is conscious (with 100% confidence) is fundamentally unknowable, since consciousness is inherently subjective (see solipsism). Regardless, some scientists strongly believe consciousness is the consequence of computational processes which are substrate-neutral. On the contrary, numerous scientists believe consciousness may be the result of some form of quantum computation dependent on substrate (see quantum mind).

In light of uncertainty on whether to regard uploads as conscious, Sandberg proposes a cautious approach:

Principle of assuming the most (PAM): Assume that any emulated system could have the same mental properties as the original system and treat it correspondingly.

It is argued that if a computational copy of one's mind did exist, it would be impossible for one to recognize it as their own mind. The argument for this stance is the following: for a computational mind to recognize an emulation of itself, it must be capable of deciding whether two Turing machines (namely, itself and the proposed emulation) are functionally equivalent. This task is uncomputable due to the undecidability of equivalence, thus there cannot exist a computational procedure in the mind that is capable of recognizing an emulation of itself.

Copying Vs. Moving

A philosophical issue with mind uploading is whether the newly generated digital mind is really the "same" sentience, or simply an exact copy with the same memories and personality. This issue is especially obvious when the original remains essentially unchanged by the procedure, thereby resulting in a copy which could potentially have rights separate from the unaltered, obvious original.

Most projected brain scanning technologies, such as serial sectioning of the brain, would necessarily be destructive, and the original brain would not survive the brain scanning procedure. But if it can be kept intact, the computer-based consciousness could be a copy of the still-living biological person. It is in that case implicit that copying a consciousness could be as feasible as literally moving it into one or several copies, since these technologies generally involve simulation of a human brain in a computer of some sort, and digital files such as computer programs can be copied precisely. It is assumed that once the versions are exposed to different sensory inputs, their experiences would begin to diverge, but all their memories up until the moment of the copying would remain the same.

The problem is made even more apparent through the possibility of creating a potentially infinite number of initially identical copies of the original person, which would of course all exist simultaneously as distinct beings with their own emotions and thoughts. The most parsimonious view of this phenomenon is that the two (or more) minds would share memories of their past but from the point of duplication would simply be distinct minds.

Toward the goal of resolving the copy-vs-move debate, some have argued for a third way of conceptualizing the process, which is described by such terms as *split* and *divergence*. The distinguishing feature of this third terminological option is that while *moving* implies that a single instance relocates in space and while *copying* invokes problematic connotations (a *copy* is often denigrated in status relative to its *original*), the notion of a *split* better illustrates that some kinds of entities might become two separate instances, but without the imbalanced associations assigned to originals and copies, and that such equality may apply to minds. Depending on computational capacity, the simulation›s subjective time may be faster or slower than elapsed physical time, resulting in that the simulated mind would perceive that the physical world is running in slow motion or fast motion respectively, while biological persons will see the simulated mind in fast or slow motion respectively.

A brain simulation can be started, paused, backed-up and rerun from a saved backup state at any time. The simulated mind would in the latter case forget everything that has happened after the instant of backup, and perhaps not even be aware that it is repeating itself. An older version of a simulated mind may meet a younger version and share experiences with it.

One proposed route for mind uploading is gradual transfer of functions from an «aging biological brain» into an exocortex.

Ethical and Legal Implications

The process of developing emulation technology raises ethical issues related to animal welfare and artificial consciousness. The neuroscience required to develop brain emulation would require animal experimentation, first on invertebrates and then on small mammals before moving on to humans. Sometimes the animals would just need to be euthanized in order to extract, slice, and scan their brains, but sometimes behavioral and *in vivo* measures would be required, which might cause pain to living animals.

In addition, the resulting animal emulations themselves might suffer, depending on one›s views about consciousness. Bancroft argues for the plausibility of consciousness in brain simulations on the basis of the «fading qualia» thought experiment of David Chalmers. He then concludes:

If, as I argue above, a sufficiently detailed computational simulation of the brain is potentially operationally equivalent to an organic brain, it follows that we must consider extending protections against suffering to simulations.

It might help reduce emulation suffering to develop virtual equivalents of anaesthesia, as well as to omit processing related to pain and/or consciousness. However, some experiments might require a fully functioning and suffering animal emulation. Animals might also suffer by accident due to flaws and lack of insight into what parts of their brains are suffering.Questions also arise regarding the moral status of *partial* brain emulations, as well as creating neuromorphic emulations that draw inspiration from biological brains but are built somewhat differently.

Brain emulations could be erased by computer viruses or malware, without need to destroy the underlying hardware. This may make assassination easier than for physical humans. The attacker might take the computing power for its own use.

Many questions arise regarding the legal personhood of emulations. Would they be given the rights of biological humans? If a person makes an emulated copy of himself and then dies, does the emulation inherit his property and official positions? Could the emulation ask to «pull the plug» when its biological version was terminally ill or in a coma? Would it help to treat emulations as adolescents for a few years so that the biological creator would maintain temporary control? Would criminal emulations receive the death penalty, or would they be given forced data modification as a form of «rehabilitation»? Could an upload have marriage and child-care rights?

If simulated minds would come true and if they were assigned rights of their own, it may be difficult to ensure the protection of «digital human rights». For example, social science researchers might be tempted to secretly expose simulated minds, or whole isolated societies of simulated minds, to controlled experiments in which many copies of the same minds are exposed (serially or simultaneously) to different test conditions.

Political and Economic Implications

Emulations could create a number of conditions that might increase risk of war, including inequality, changes of power dynamics, a possible technological arms race to build emulations first, first-strike advantages, strong loyalty and willingness to "die" among emulations, and triggers for racist, xenophobic, and religious prejudice. If emulations run much faster than humans, there might not be enough time for human leaders to make wise decisions or negotiate. It's possible that humans would react violently against growing power of emulations, especially if they depress human wages. Or maybe emulations wouldn't trust each other, and even well intentioned defensive measures might be interpreted as offense.

Emulation Timelines and AI Risk

There are very few feasible technologies that humans have refrained from developing. The neuroscience and computer-hardware technologies that may make brain emulation possible are widely desired for other reasons, so cutting off funding doesn't seem to be an option. If we assume that emulation technology will arrive, a question becomes whether we should accelerate or slow its advance.

Arguments for speeding up brain-emulation research:

- If neuroscience is the bottleneck on brain emulation rather than computing power, emulation advances may be more erratic and unpredictable based on when new scientific discoveries happen. Limited computing power would mean the first emulations would run slower and so would be easier to adapt to, and there would be more time for the technology to transition through society.
- Improvements in manufacturing, 3D printing, and nanotechnology may accelerate hardware production, which could increase the "computing overhang" from excess hardware relative to neuroscience.
- If one AI-development group had a lead in emulation technology, it would have more subjective time to win an arms race to build the first superhuman AI. Because it would be less rushed, it would have more freedom to consider AI risks.

Arguments for slowing down brain-emulation research:

- Greater investment in brain emulation and associated cognitive science might enhance the ability of artificial intelligence (AI) researchers to create "neuromorphic" (brain-inspired) algorithms, such as neural networks, reinforcement learning, and hierarchical perception. This could accelerate risks from uncontrolled AI. Participants at a 2011 AI workshop estimated an 85% probability that neuromorphic AI would arrive before brain emulation. This was based on the idea that brain emulation would require understanding some brain components, and it would be easier to tinker with these than to reconstruct the entire brain in its original form. By a very narrow margin, the participants on balance leaned toward the view that accelerating brain emulation would increase expected AI risk.
- Waiting might give society more time to think about the consequences of brain emulation and develop institutions to improve cooperation.

Emulation research would also speed up neuroscience as a whole, which might accelerate medical advances, cognitive enhancement, lie detectors, and capability for psychological manipulation.

Emulations might be easier to control than *de novo* AI because

- We understand better human abilities, behavioral tendencies, and vulnerabilities, so control measures might be more intuitive and easier to plan for.
- Emulations could more easily inherit human motivations.
- Emulations are harder to manipulate than *de novo* AI, because brains are messy and complicated; this could reduce risks of their rapid takeoff. Also, emulations

may be bulkier and require more hardware than AI, which would also slow the speed of a transition. Unlike AI, an emulation wouldn't be able to rapidly expand beyond the size of a human brain. Emulations running at digital speeds would have less intelligence differential vis-à-vis AI and so might more easily control AI.

As counterpoint to these considerations, Bostrom notes some downsides:

- Even if we better understand human behavior, the *evolution* of emulation behavior under self-improvement might be much less predictable than the evolution of safe *de novo* AI under self-improvement.
- Emulations may not inherit all human motivations. Perhaps they would inherit our darker motivations or would behave abnormally in the unfamiliar environment of cyberspace.
- Even if there›s a slow takeoff toward emulations, there would still be a second transition to *de novo* AI later on. Two intelligence explosions may mean more total risk.

Advocates

Ray Kurzweil, director of engineering at Google, claims to know and foresee that people will be able to "upload" their entire brains to computers and become "digitally immortal" by 2045. Kurzweil made this claim for many years, e.g. during his speech in 2013 at the Global Futures 2045 International Congress in New York, which claims to subscribe to a similar set of beliefs. Mind uploading is also advocated by a number of researchers in neuroscience and artificial intelligence, such as Marvin Minsky while he was still alive. In 1993, Joe Strout created a small web site called the Mind Uploading Home Page, and began advocating the idea in cryonics circles and elsewhere on the net. That site has not been actively updated in recent years, but it has spawned other sites including MindUploading.org, run by Randal A. Koene, who also moderates a mailing list on the topic. These advocates see mind uploading as a medical procedure which could eventually save countless lives.

Many transhumanists look forward to the development and deployment of mind uploading technology, with transhumanists such as Nick Bostrom predicting that it will become possible within the 21st century due to technological trends such as Moore's law.

Michio Kaku, in collaboration with Science, hosted a documentary, *Sci Fi Science: Physics of the Impossible*, based on his book *Physics of the Impossible*. Episode four, titled "How to Teleport", mentions that mind uploading *via* techniques such as quantum entanglement and whole brain emulation using an advanced MRI machine may enable people to be transported to vast distances at near light-speed.

The book *Beyond Humanity: CyberEvolution and Future Minds* by Gregory S. Paul & Earl D. Cox, is about the eventual (and, to the authors, almost inevitable) evolution of computers into sentient beings, but also deals with human mind transfer. Richard Doyle's *Wetwares: Experiments in PostVital Living* deals extensively with uploading from the perspective of distributed embodiment, arguing for example that humans are currently part of the "artificial life phenotype". Doyle's vision reverses the polarity on uploading, with artificial life forms such as uploads actively seeking out biological embodiment as part of their reproductive strategy.

Skeptics

Kenneth D. Miller, a professor of neuroscience at Columbia and a co-director of the Center for Theoretical Neuroscience, raised doubts about the practicality of mind uploading.

NEURAL CODING

Neural coding is a neuroscience related field concerned with characterizing the relationship between the stimulus and the individual or ensemble neuronal responses and the relationship among the electrical activity of the neurons in the ensemble. Based on the theory that sensory and other information is represented in the brain by networks of neurons, it is thought that neurons can encode both digital and analog information.

Overview

Neurons are remarkable among the cells of the body in their ability to propagate signals rapidly over large distances. They do this by generating characteristic electrical pulses called action potentials: voltage spikes that can travel down nerve fibers. Sensory neurons change their activities by firing sequences of action potentials in various temporal patterns, with the presence of external sensory stimuli, such as light, sound, taste, smell and touch. It is known that information about the stimulus is encoded in this pattern of action potentials and transmitted into and around the brain.

Although action potentials can vary somewhat in duration, amplitude and shape, they are typically treated as identical stereotyped events in neural coding studies. If the brief duration of an action potential (about 1ms) is ignored, an action potential sequence, or spike train, can be characterized simply by a series of all-or-none point events in time. The lengths of interspike intervals (ISIs) between two successive spikes in a spike train often vary, apparently randomly. The study of neural coding involves measuring and characterizing how stimulus attributes, such as light or sound intensity, or motor actions, such as the direction of an arm movement, are represented by neuron action potentials or spikes. In order to describe and analyze neuronal firing, statistical methods and methods of probability theory and stochastic point processes have been widely applied.

With the development of large-scale neural recording and decoding technologies, researchers have begun to crack the neural code and already provided the first glimpse into the real-time neural code as memory is formed and recalled in the hippocampus, a brain region known to be central for memory formation. Neuroscientists have initiated several large-scale brain decoding projects.

Encoding and Decoding

The link between stimulus and response can be studied from two opposite points of view. Neural encoding refers to the map from stimulus to response. The main focus is to understand how neurons respond to a wide variety of stimuli, and to construct models that attempt to predict responses to other stimuli. Neural decoding refers to the reverse map, from response to stimulus, and the challenge is to reconstruct a stimulus, or certain aspects of that stimulus, from the spike sequences it evokes.

Coding schemes

A sequence, or 'train', of spikes may contain information based on different coding schemes. In motor neurons, for example, the strength at which an innervated muscle is flexed depends solely on the 'firing rate', the average number of spikes per unit time (a 'rate code'). At the other end, a complex 'temporal code' is based on the precise timing of single spikes. They may be locked to an external stimulus such as in the visual and auditory system or be generated intrinsically by the neural circuitry.

Whether neurons use rate coding or temporal coding is a topic of intense debate within the neuroscience community, even though there is no clear definition of what these terms mean. In one theory, termed "neuroelectrodynamics", the following coding schemes are all considered to be epiphenomena, replaced instead by molecular changes reflecting the spatial distribution of electric fields within neurons as a result of the broad electromagnetic spectrum of action potentials, and manifested in information as spike directivity.

Rate Coding

The rate coding model of neuronal firing communication states that as the intensity of a stimulus increases, the frequency or rate of action potentials, or "spike firing", increases. Rate coding is sometimes called frequency coding.

Rate coding is a traditional coding scheme, assuming that most, if not all, information about the stimulus is contained in the firing rate of the neuron. Because the sequence of action potentials generated by a given stimulus varies from trial to trial, neuronal responses are typically treated statistically or probabilistically. They may be characterized by firing rates, rather than as specific spike sequences. In most sensory systems, the firing rate increases, generally non-linearly, with increasing stimulus intensity. Any information possibly encoded in the temporal structure of the spike train is ignored. Consequently, rate coding is inefficient but highly robust with respect to the ISI 'noise'.

During rate coding, precisely calculating firing rate is very important. In fact, the term "firing rate" has a few different definitions, which refer to different averaging procedures, such as an average over time or an average over several repetitions of experiment.

In rate coding, learning is based on activity-dependent synaptic weight modifications.

Rate coding was originally shown by ED Adrian and Y Zotterman in 1926. In this simple experiment different weights were hung from a muscle. As the weight of the stimulus increased, the number of spikes recorded from sensory nerves innervating the muscle also increased. From these original experiments, Adrian and Zotterman concluded that action potentials were unitary events, and that the frequency of events, and not individual event magnitude, was the basis for most inter-neuronal communication.

In the following decades, measurement of firing rates became a standard tool for describing the properties of all types of sensory or cortical neurons, partly due to the relative ease of measuring rates experimentally. However, this approach neglects all the information possibly contained in the exact timing of the spikes. During recent years, more and more experimental evidence has suggested that a straightforward firing rate concept based on temporal averaging may be too simplistic to describe brain activity.

Spike-Count Rate

The Spike-count rate, also referred to as temporal average, is obtained by counting the number of spikes that appear during a trial and dividing by the duration of trial. The length T of the time window is set by experimenter and depends on the type of neuron recorded from and the stimulus. In practice, to get sensible averages, several spikes should occur within the time window. Typical values are T = 100 ms or T = 500 ms, but the duration may also be longer or shorter. The spike-count rate can be determined from a single trial, but at the expense of losing all temporal resolution about variations in neural response during the course of the trial. Temporal averaging can work well in cases where the stimulus is constant or slowly varying and does not require a fast reaction of the organism — and this is the situation usually encountered in experimental protocols. Real-world input, however, is hardly stationary, but often changing on a fast time scale. For example, even when viewing a static image, humans perform saccades, rapid changes of the direction of gaze. The image projected onto the retinal photoreceptors changes therefore every few hundred milliseconds. Despite its shortcomings, the concept of a spike-count rate code is widely used not only in experiments, but also in models of neural networks. It has led to the idea that a neuron transforms information about a single input variable (the stimulus strength) into a single continuous output variable (the firing rate). There is a growing body of evidence that in Purkinje neurons, at least, information is not simply encoded in firing but also in the timing and duration of non-firing, quiescent periods.

Time-Dependent Firing Rate

The time-dependent firing rate is defined as the average number of spikes (averaged over trials) appearing during a short interval between times t and t+Δt, divided by the duration of the interval. It works for stationary as well as for time-dependent stimuli. To experimentally measure the time-dependent firing rate, the experimenter records from a neuron while stimulating with some input sequence. The same stimulation sequence is repeated several times and the neuronal response is reported in a Peri-Stimulus-Time Histogram (PSTH). The time t is measured with respect to the start of the stimulation sequence. The Δt must be large enough (typically in the range of one or a few milliseconds) so there are sufficient number of spikes within the interval to obtain a reliable estimate of the average. The number of occurrences of spikes $n_K(t;t+\Delta t)$ summed over all repetitions of the experiment divided by the number K of repetitions is a measure of the typical activity of the neuron between time t and t+Δt. A further division by the interval length Δt yields time-dependent firing rate r(t) of the neuron, which is equivalent to the spike density of PSTH. For sufficiently small Δt, r(t)Δt is the average number of spikes occurring between times t and t+Δt over multiple trials. If Δt is small, there will never be more than one spike within the interval between t and t+Δt on any given trial. This means that r(t)Δt is also the fraction of trials on which a spike occurred between those times. Equivalently, r(t)Δt is the probability that a spike occurs during this time interval.

As an experimental procedure, the time-dependent firing rate measure is a useful method to evaluate neuronal activity, in particular in the case of time-dependent stimuli. The obvious problem with this approach is that it can not be the coding scheme used by neurons in the brain. Neurons can not wait for the stimuli to repeatedly present in an exactly same manner before generating response. Nevertheless, the experimental time-dependent firing rate measure can make sense, if there are large populations of independent neurons that receive the same stimulus. Instead of recording from a population of N neurons in a single run, it is experimentally easier to record from a single neuron and average over N repeated runs. Thus, the time-dependent firing rate coding relies on the implicit assumption that there are always populations of neurons.

Temporal Coding

When precise spike timing or high-frequency firing-rate fluctuations are found to carry information, the neural code is often identified as a temporal code. A number of studies have found that the temporal resolution of the neural code is on a millisecond time scale, indicating that precise spike timing is a significant element in neural coding.

Neurons exhibit high-frequency fluctuations of firing-rates which could be noise or could carry information. Rate coding models suggest that these irregularities are noise, while temporal coding models suggest that they encode information. If the nervous system only used rate codes to convey information, a more consistent, regular firing rate would have been evolutionarily advantageous, and neurons would have utilized this code over other less robust options. Temporal coding supplies an alternate explanation for the "noise," suggesting that it actually encodes information and affects neural processing. To model this idea, binary symbols can be used to mark the spikes: 1 for a spike, 0 for no spike. Temporal coding allows the sequence 000111000111 to mean something different from 001100110011, even though the mean firing rate is the same for both sequences, at 6 spikes/10 ms. Until recently, scientists had put the most emphasis on rate encoding as an explanation for post-synaptic potential patterns. However, functions of the brain are more temporally precise than the use of only rate encoding seems to allow. In other words, essential information could be lost due to the inability of the rate code to capture all the available information of the spike train. In addition, responses are different enough between similar (but not identical) stimuli to suggest that the distinct patterns of spikes contain a higher volume of information than is possible to include in a rate code.

Temporal codes employ those features of the spiking activity that cannot be described by the firing rate. For example, time to first spike after the stimulus onset, characteristics based on the second and higher statistical moments of the ISI probability distribution, spike randomness, or precisely timed groups of spikes (temporal patterns) are candidates for temporal codes. As there is no absolute time reference in the nervous system, the information is carried either in terms of the relative timing of spikes in a population of neurons or with respect to an ongoing brain oscillation. The temporal structure of a spike train or firing rate evoked by a stimulus is determined both by the dynamics of the stimulus and by the nature of the neural encoding process. Stimuli that change rapidly tend to generate precisely timed spikes and rapidly changing firing rates no matter what neural coding strategy is being used. Temporal coding refers to temporal precision in the response that does not arise solely from the dynamics of the stimulus, but that nevertheless relates to properties of the stimulus. The interplay between stimulus and encoding dynamics makes the identification of a temporal code difficult. In temporal coding, learning can be explained by activity-dependent synaptic delay modifications. The modifications can themselves depend not only on spike rates (rate coding) but also on spike timing patterns (temporal coding), i.e., can be a special case of spike-timing-dependent plasticity. The issue of temporal coding is distinct and independent from the issue of independent-spike coding. If each spike is independent of all the other spikes in the train, the temporal character of the neural code is determined by the behavior of time-dependent firing rate r(t). If r(t) varies slowly with time, the code is typically called a rate code, and if it varies rapidly, the code is called temporal.

Temporal Coding in Sensory Systems

For very brief stimuli, a neuron's maximum firing rate may not be fast enough to produce more than a single spike. Due to the density of information about the abbreviated stimulus contained in this single spike, it would seem that the timing of the spike itself would have to convey more information than simply the average frequency of action potentials over a given period of time. This model is especially important for sound localization, which occurs within the brain on the order of milliseconds. The brain must obtain a large quantity of information based on a relatively short neural response. Additionally, if low firing rates on the order of ten spikes per second must be distinguished from arbitrarily close rate coding for different stimuli, then a neuron trying to discriminate these two stimuli may need to wait for a second or more to accumulate enough information. This is not consistent with numerous organisms which are able to discriminate between stimuli in the time frame of milliseconds, suggesting that a rate code is not the only model at work. To account for the fast encoding of visual stimuli, it has been suggested that neurons of the retina encode visual information in the latency time between stimulus onset and first action potential, also called latency to first spike. This type of temporal coding has been shown also in the auditory and somato-sensory system. The main drawback of such a coding scheme is its sensitivity to intrinsic neuronal fluctuations. In the primary visual cortex of macaques, the timing of the first spike relative to the start of the stimulus was found to provide more information than the interval between spikes. However, the interspike interval could be used to encode additional information, which is especially important when the spike rate reaches its limit, as in high-contrast situations. For this reason, temporal coding may play a part in coding defined edges rather than gradual transitions.

The mammalian gustatory system is useful for studying temporal coding because of its fairly distinct stimuli and the easily discernible responses of the organism. Temporally encoded information may help an organism discriminate between different tastants of the same category (sweet, bitter, sour, salty, umami) that elicit very similar responses in terms of spike count. The temporal component of the pattern elicited by each tastant may be used to determine its identity (e.g., the difference between two bitter tastants, such as quinine and denatonium). In this way, both rate coding and temporal coding may be used in the gustatory system – rate for basic tastant type, temporal for more specific differentiation. Research on mammalian gustatory system has shown that there is an abundance of information present in temporal patterns across populations of neurons, and this information is different from that which is determined by rate coding schemes. Groups of neurons may synchronize in response to a stimulus. In studies dealing with the front cortical portion of the brain in primates, precise patterns with short time scales only a few milliseconds in length were found across small populations of neurons which correlated with certain information processing behaviors. However, little information could be determined from the patterns; one possible theory is they represented the higher-order processing taking place in the brain.

As with the visual system, in mitral/tufted cells in the olfactory bulb of mice, first-spike latency relative to the start of a sniffing action seemed to encode much of the information about an odor. This strategy of using spike latency allows for rapid identification of and reaction to an odorant. In addition, some mitral/tufted cells have specific firing patterns for given odorants. This type of extra information could help in recognizing a certain odor, but is not completely necessary, as average spike count over the course of the animal's sniffing was also a good identifier. Along the same lines, experiments done with the olfactory system of rabbits showed distinct patterns which correlated with different subsets of odorants, and a similar result was obtained in experiments with the locust olfactory system.

Temporal Coding Applications

The specificity of temporal coding requires highly refined technology to measure informative, reliable, experimental data. Advances made in optogenetics allow neurologists to control spikes in individual neurons, offering electrical and spatial single-cell resolution. For example, blue light causes the light-gated ion channel channelrhodopsin to open, depolarizing the cell and producing a spike. When blue light is not sensed by the cell, the channel closes, and the neuron ceases to spike. The pattern of the spikes matches the pattern of the blue light stimuli. By inserting channelrhodopsin gene sequences into mouse DNA, researchers can control spikes and therefore certain behaviors of the mouse (e.g., making the mouse turn left).Researchers, through optogenetics, have the tools to effect different temporal codes in a neuron while maintaining the same mean firing rate, and thereby can test whether or not temporal coding occurs in specific neural circuits. Optogenetic technology also has the potential to enable the correction of spike abnormalities at the root of several neurological and psychological disorders. If neurons do encode information in individual spike timing patterns, key signals could be missed by attempting to crack the code while looking only at mean firing rates. Understanding any temporally encoded aspects of the neural code and replicating these sequences in neurons could allow for greater control and treatment of neurological disorders such as depression, schizophrenia, and Parkinson's disease. Regulation of spike intervals in single cells more precisely controls brain activity than the addition of pharmacological agents intravenously.

Phase-of-Firing Code

Phase-of-firing code is a neural coding scheme that combines the spike count code with a time reference based on oscillations. This type of code takes into account a time label for each spike according to a time reference based on phase of local ongoing oscillations at low or high frequencies. A feature of this code is that neurons adhere to a preferred order of spiking, resulting in firing sequence. It has been shown that neurons in some cortical sensory areas encode rich naturalistic stimuli in terms of their spike times relative to the phase of ongoing network fluctuations, rather than only in terms of their spike count. Oscillations reflect local field potential signals. It is often categorized as a temporal code although the time label used for spikes is coarse grained. That is, four discrete values for phase are enough to represent all the information content in this kind of code with respect to the phase of oscillations in low frequencies. Phase-of-firing code is loosely based on the phase precession phenomena observed in place cells of the hippocampus. (Also see Phase resetting in neurons) Phase code has been shown in visual cortex to involve also high-frequency oscillations. Within a cycle of gamma oscillation, each neuron has its own preferred relative firing time. As a result, an entire population of neurons generates a firing sequence that has a duration of up to about 15 ms.

Population Coding

Population coding is a method to represent stimuli by using the joint activities of a number of neurons. In population coding, each neuron has a distribution of responses over some set of inputs, and the responses of many neurons may be combined to determine some value about the inputs.

From the theoretical point of view, population coding is one of a few mathematically well-formulated problems in neuroscience. It grasps the essential features of neural coding and yet is simple enough for theoretic analysis. Experimental studies have revealed that this coding paradigm is widely used in the sensor and motor areas of the brain. For example, in the visual area medial temporal (MT), neurons are tuned to the moving direction. In response to an object moving in a particular direction, many neurons in MT fire with a noise-corrupted and bell-shaped activity pattern across the population. The moving direction of the object is retrieved from the population activity, to be immune from the fluctuation existing in a single neuron's signal. In one classic example in the primary motor cortex, Apostolos Georgopoulos and colleagues trained monkeys to move a joystick towards a lit target. They found that a single neuron would fire for multiple target directions. However it would fire fastest for one direction and more slowly depending on how close the target was to the neuron's 'preferred' direction.

Kenneth Johnson originally derived that if each neuron represents movement in its preferred direction, and the vector sum of all neurons is calculated (each neuron has a firing rate and a preferred direction), the sum points in the direction of motion. In this manner, the population of neurons codes the signal for the motion. This particular population code is referred to as population vector coding. This particular study divided the field of motor physiologists between Evarts' "upper motor neuron" group, which followed the hypothesis that motor cortex neurons contributed to control of single muscles, and the Georgopoulos group studying the representation of movement directions in cortex.

The Johns Hopkins University Neural Encoding laboratory led by Murray Sachs and Eric Young developed place-time population codes, termed the Averaged-Localized-

Synchronized-Response (ALSR) code for neural representation of auditory acoustic stimuli. This exploits both the place or tuning within the auditory nerve, as well as the phase-locking within each nerve fiber Auditory nerve. The first ALSR representation was for steady-state vowels; ALSR representations of pitch and formant frequencies in complex, non-steady state stimuli were demonstrated for voiced-pitch and formant representations in consonant-vowel syllables. The advantage of such representations is that global features such as pitch or formant transition profiles can be represented as global features across the entire nerve simultaneously via both rate and place coding.

Population coding has a number of other advantages as well, including reduction of uncertainty due to neuronal variability and the ability to represent a number of different stimulus attributes simultaneously. Population coding is also much faster than rate coding and can reflect changes in the stimulus conditions nearly instantaneously. Individual neurons in such a population typically have different but overlapping selectivities, so that many neurons, but not necessarily all, respond to a given stimulus.

Typically an encoding function has a peak value such that activity of the neuron is greatest if the perceptual value is close to the peak value, and becomes reduced accordingly for values less close to the peak value. It follows that the actual perceived value can be reconstructed from the overall pattern of activity in the set of neurons. The Johnson/Georgopoulos vector coding is an example of simple averaging. A more sophisticated mathematical technique for performing such a reconstruction is the method of maximum likelihood based on a multivariate distribution of the neuronal responses. These models can assume independence, second order correlations , or even more detailed dependencies such as higher order maximum entropy models or copulas.

Correlation Coding

The correlation coding model of neuronal firing claims that correlations between action potentials, or "spikes", within a spike train may carry additional information above and beyond the simple timing of the spikes. Early work suggested that correlation between spike trains can only reduce, and never increase, the total mutual information present in the two spike trains about a stimulus feature. However, this was later demonstrated to be incorrect. Correlation structure can increase information content if noise and signal correlations are of opposite sign. Correlations can also carry information not present in the average firing rate of two pairs of neurons. A good example of this exists in the pentobarbital-anesthetized marmoset auditory cortex, in which a pure tone causes an increase in the number of correlated spikes, but not an increase in the mean firing rate, of pairs of neurons.

Independent-Spike Coding

The independent-spike coding model of neuronal firing claims that each individual action potential, or "spike", is independent of each other spike within the spike train.

Position Coding

Plot of typical position coding

A typical population code involves neurons with a Gaussian tuning curve whose means vary linearly with the stimulus intensity, meaning that the neuron responds most strongly (in terms of spikes per second) to a stimulus near the mean. The actual intensity could be recovered as the stimulus level corresponding to the mean of the neuron with the greatest response. However, the noise inherent in neural responses means that a maximum likelihood estimation function is more accurate.

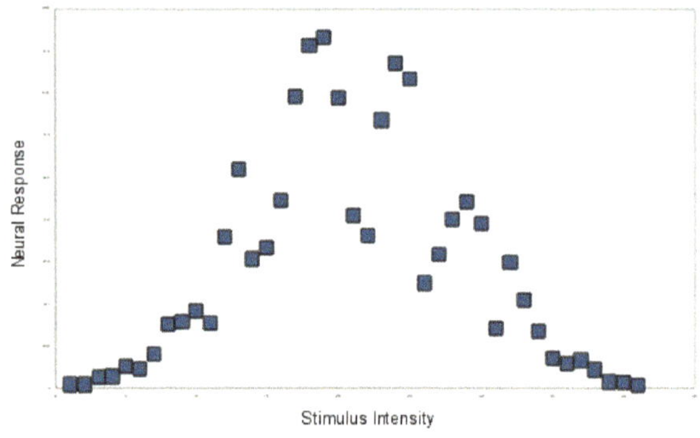

Neural responses are noisy and unreliable.

This type of code is used to encode continuous variables such as joint position, eye position, color, or sound frequency. Any individual neuron is too noisy to faithfully encode the variable using rate coding, but an entire population ensures greater fidelity and precision. For a population of unimodal tuning curves, i.e. with a single peak, the precision typically scales linearly with the number of neurons. Hence, for half the precision, half as many neurons are required. In contrast, when the tuning curves have multiple peaks, as in grid cells that represent space, the precision of the population can scale exponentially with the number of neurons. This greatly reduces the number of neurons required for the same precision.

Sparse Coding

The sparse code is when each item is encoded by the strong activation of a relatively small set of neurons. For each item to be encoded, this is a different subset of all available neurons. As a consequence, sparseness may be focused on temporal sparseness ("a relatively small number of time periods are active") or on the sparseness in an activated population of neurons. In this latter case, this may be defined in one time period as the number of activated neurons relative to the total number of neurons in the population. This seems to be a hallmark of neural computations since compared to traditional computers, information is massively distributed across neurons. A major result in neural coding from Olshausen and Field is that sparse coding of natural images produces wavelet-like oriented filters that resemble the receptive fields of simple cells in the visual cortex. The capacity of sparse codes may be increased by simultaneous use of temporal coding, as found in the locust olfactory system.

Given a potentially large set of input patterns, sparse coding algorithms (e.g. Sparse Autoencoder) attempt to automatically find a small number of representative patterns which, when combined in the right proportions, reproduce the original input patterns. The sparse coding for the input then consists of those representative patterns. For example, the very large set of English sentences can be encoded by a small number of symbols (i.e. letters, numbers, punctuation, and spaces) combined in a particular order for a particular sentence, and so a sparse coding for English would be those symbols.

Linear Generative Model

Most models of sparse coding are based on the linear generative model. In this model, the symbols are combined in a linear fashion to approximate the input.

More formally, given a k-dimensional set of real-numbered input vectors $\vec{\xi} \in \mathbb{R}^k$ the goal of sparse coding is to determine n k-dimensional basis vectors $\vec{b_1}, \ldots, \vec{b_n} \in \mathbb{R}^k$ along with a sparse n-dimensional vector of weights or coefficients $\vec{s} \in \mathbb{R}^n$ for each input vector, so that a linear combination of the basis vectors with proportions given by the coefficients results in a close approximation to the input vector. $\vec{\xi} \approx \sum_{j=1}^{n} s_j \vec{b}_j$.

The codings generated by algorithms implementing a linear generative model can be classified into codings with *soft sparseness* and those with *hard sparseness*. These refer to the distribution of basis vector coefficients for typical inputs. A coding with soft sparseness has a smooth Gaussian-like distribution, but peakier than Gaussian, with many zero values, some small absolute values, fewer larger absolute values, and very few very large absolute values. Thus, many of the basis vectors are active. Hard sparseness, on the other hand, indicates that there are many zero values, *no* or *hardly any* small absolute values, fewer larger absolute values, and very few very large absolute values, and thus few of the basis vectors are active. This is appealing from a metabolic perspective: less energy is used when fewer neurons are firing.

Another measure of coding is whether it is *critically complete* or *overcomplete*. If the number of basis vectors n is equal to the dimensionality k of the input set, the coding is said to be critically complete. In this case, smooth changes in the input vector result in abrupt changes in the coefficients, and the coding is not able to gracefully handle small scalings, small translations, or noise in the inputs. If, however, the number of basis vectors is larger than the dimensionality of the input set, the coding is *overcomplete*. Overcomplete codings smoothly interpolate between input vectors and are robust under input noise. The human primary visual cortex is estimated to be overcomplete by a factor of 500, so that, for example, a 14 x 14 patch of input (a 196-dimensional space) is coded by roughly 100,000 neurons.

Biological Evidence

Sparse coding may be a general strategy of neural systems to augment memory capacity. To adapt to their environments, animals must learn which stimuli are associated with rewards or punishments and distinguish these reinforced stimuli from similar but irrelevant ones. Such task requires implementing stimulus-specific associative memories in which only a few neurons out of a population respond to any given stimulus and each neuron responds to only a few stimuli out of all possible stimuli.

Theoretical work on Sparse distributed memory has suggested that sparse coding increases the capacity of associative memory by reducing overlap between representations. Experimentally, sparse representations of sensory information have been observed in many systems, including vision, audition, touch, and olfaction. However, despite the accumulating evidence for widespread sparse coding and theoretical arguments for its importance, a demonstration that sparse coding improves the stimulus-specificity of associative memory has been lacking until recently.

Some progress has been made in 2014 by Gero Miesenböck's lab at the University of Oxford analyzing Drosophila Olfactory system. In Drosophila, sparse odor coding by the Kenyon cells of the mushroom body is thought to generate a large number of precisely addressable locations for the storage of odor-specific memories. Lin et al. demonstrated that sparseness is controlled by a negative feedback circuit between Kenyon cells and the GABAergic anterior paired lateral (APL) neuron. Systematic activation and blockade of each leg of this feedback circuit show that Kenyon cells activate APL and APL inhibits Kenyon cells. Disrupting the Kenyon cell-APL feedback loop decreases the sparseness of Kenyon cell odor responses, increases inter-odor correlations, and prevents flies from learning to discriminate similar, but not dissimilar, odors. These results suggest that feedback inhibition suppresses Kenyon cell activity to maintain sparse, decorrelated odor coding and thus the odor-specificity of memories.

REFERENCES

1. Vidal, JJ (1973). "Toward direct brain-computer communication". Annual Review of Biophysics and Bioengineering. 2 (1): 157–80. doi:10.1146/annurev.bb.02.060173.001105. PMID 4583653.
2. J. Vidal (1977). "Real-Time Detection of Brain Events in EEG" (PDF). IEEE Proceedings. 65 (5): 633–641. doi:10.1109/PROC.1977.10542.
3. Levine, SP; Huggins, JE; Bement, SL; Kushwaha, RK; Schuh, LA; Rohde, MM; Passaro, EA; Ross, DA; Elisevich, KV; et al. (2000). "A direct brain

interface based on event-related potentials". IEEE transactions on rehabilitation engineering : a publication of the IEEE Engineering in Medicine and Biology Society. 8 (2): 180–5. doi:10.1109/86.847809. PMID 10896180.

4. Wolpaw, J.R. and Wolpaw, E.W. (2012). "Brain-Computer Interfaces: Something New Under the Sun". In: Brain-Computer Interfaces: Principles and Practice, Wolpaw, J.R. and Wolpaw (eds.), E.W. Oxford University Press.
5. Wolpaw J.R.; Birbaumer N.; McFarland D.J.; Pfurtscheller G.; Vaughan T. M. (2002). "Brain–computer interfaces for communication and control". Clinical Neurophysiology. 113 (6): 767–791. doi:10.1016/s1388-2457(02)00057-3.
6. Allison B.Z.; Wolpaw E.W.; Wolpaw J.R. (2007). "Brain computer interface systems: Progress and prospects". British review of medical devices. 4 (4): 463–474. doi:10.1586/17434440.4.4.463.
7. NIH Publication No. 11-4798 (1 March 2011). "Cochlear Implants". National Institute on Deafness and Other Communication Disorders.
8. Miguel Nicolelis et al. (2001) Duke neurobiologist has developed system that allows monkeys to control robot arms via brain signals Archived 19 December 2008 at the Wayback Machine.
9. Baum, Michele (6 September 2008). "Monkey Uses Brain Power to Feed Itself With Robotic Arm". Pitt Chronicle. Retrieved 6 July 2009.
10. Fetz, E. E. (1969). "Operant Conditioning of Cortical Unit Activity". Science. 163 (3870): 955–8. Bibcode:1969Sci...163..955F. doi:10.1126/science.163.3870.955. PMID 4974291.
11. Schmidt, EM; McIntosh, JS; Durelli, L; Bak, MJ (1978). "Fine control of operantly conditioned firing patterns of cortical neurons". Experimental neurology. 61 (2): 349–69. doi:10.1016/0014-4886(78)90252-2. PMID 101388.
12. Georgopoulos, A.; Lurito, J.; Petrides, M; Schwartz, A.; Massey, J. (1989). "Mental rotation of the neuronal population vector". Science. 243 (4888): 234–6. Bibcode:1989Sci...243..234G. doi:10.1126/science.2911737. PMID 2911737.
13. Lebedev, MA; Nicolelis, MA (2006). "Brain-machine interfaces: past, present and future" (PDF). Trends in Neurosciences. 29 (9): 536–46. doi:10.1016/j.tins.2006.07.004. PMID 16859758.
14. Stanley, GB; Li, FF; Dan, Y (1999). "Reconstruction of natural scenes from ensemble responses in the lateral geniculate nucleus" (PDF). Journal of Neuroscience. 19 (18): 8036–42. PMID 10479703.
15. Nicolelis, Miguel A. L.; Wessberg, Johan; Stambaugh, Christopher R.; Kralik, Jerald D.; Beck, Pamela D.; Laubach, Mark; Chapin, John K.; Kim, Jung; Biggs, S. James; et al. (2000). "Real-time prediction of hand trajectory by ensembles of cortical neurons in primates". Nature. 408 (6810): 361–5. doi:10.1038/35042582. PMID 11099043.
16. Carmena, JM; Lebedev, MA; Crist, RE; O'Doherty, JE; Santucci, DM; Dimitrov, DF; Patil, PG; Henriquez, CS; Nicolelis, MA (2003). "Learning to control a brain-machine interface for reaching and grasping by primates". PLoS Biology.

1 (2): E42. doi:10.1371/journal.pbio.0000042. PMC 261882 . PMID 14624244.

17. Lebedev, M. A.; Carmena, JM; O'Doherty, JE; Zacksenhouse, M; Henriquez, CS; Principe, JC; Nicolelis, MA (2005). "Cortical Ensemble Adaptation to Represent Velocity of an Artificial Actuator Controlled by a Brain-Machine Interface". Journal of Neuroscience. 25 (19): 4681–93. doi:10.1523/JNEUROSCI.4088-04.2005. PMID 15888644.
18. O'Doherty, JE; Lebedev, MA; Ifft, PJ; Zhuang, KZ; Shokur, S; Bleuler, H; Nicolelis, MA (2011). "Active tactile exploration using a brain–machine–brain interface". Nature. 479 (7372): 228–231. Bibcode:2011Natur.479..228O. doi:10.1038/nature10489. PMC 3236080 . PMID 21976021.
19. Serruya, MD; Hatsopoulos, NG; Paninski, L; Fellows, MR; Donoghue, JP (2002). "Instant neural control of a movement signal". Nature. 416 (6877): 141–2. Bibcode:2002Natur.416..141S. doi:10.1038/416141a. PMID 11894084.
20. Taylor, D. M.; Tillery, SI; Schwartz, AB (2002). "Direct Cortical Control of 3D Neuroprosthetic Devices". Science. 296 (5574): 1829–32. Bibcode:2002Sci...296.1829T. doi:10.1126/science.1070291. PMID 12052948.
21. Pitt team to build on brain-controlled arm Archived 4 July 2007 at the Wayback Machine., Pittsburgh Tribune Review, 5 September 2006.
22. Video on YouTube
23. Velliste, M; Perel, S; Spalding, MC; Whitford, AS; Schwartz, AB (2008). "Cortical control of a prosthetic arm for self-feeding". Nature. 453 (7198): 1098–101. Bibcode:2008Natur.453.1098V. doi:10.1038/nature06996. PMID 18509337.
24. Musallam, S.; Corneil, BD; Greger, B; Scherberger, H; Andersen, RA (2004). "Cognitive Control Signals for Neural Prosthetics". Science. 305 (5681): 258–62. Bibcode:2004Sci...305..258M. doi:10.1126/science.1097938. PMID 15247483.
25. Santucci, David M.; Kralik, Jerald D.; Lebedev, Mikhail A.; Nicolelis, Miguel A. L. (2005). "Frontal and parietal cortical ensembles predict single-trial muscle activity during reaching movements in primates". European Journal of Neuroscience. 22 (6): 1529–40. doi:10.1111/j.1460-9568.2005.04320.x. PMID 16190906.
26. Huber, D; Petreanu, L; Ghitani, N; Ranade, S; Hromádka, T; Mainen, Z; Svoboda, K (2008). "Sparse optical microstimulation in barrel cortex drives learned behaviour in freely moving mice". Nature. 451 (7174): 61–4. Bibcode:2008Natur.451...61H. doi:10.1038/nature06445. PMC 3425380 . PMID 18094685.
27. Nicolelis Miguel A. L; Lebedev Mikhail A (2009). "Principles of Neural Ensemble Physiology Underlying the Operation of Brain-Machine Interfaces". Nature Reviews Neuroscience. 10 (7): 530–540. doi:10.1038/nrn2653.
28. Zander, Thorsten O; Kothe, Christian. "Towards passive brain–computer interfaces: applying brain–computer interface technology to human–machine systems in general". Journal of Neural Engineering. 8 (2): 025005. doi:10.1088/1741-2560/8/2/025005.

29. "The Annual BCI Research Award 2014 – The Winners". Biosignal.at. 15 June 2011. Retrieved 19 December 2016.

30. Polikov, Vadim S., Patrick A. Tresco, and William M. Reichert (2005). "Response of brain tissue to chronically implanted neural electrodes". Journal of neuroscience methods. 148 (1): 1–18.

31. Vision quest, Wired Magazine, September 2002

32. Naumann, J. Search for Paradise: A Patient's Account of the Artificial Vision Experiment (2012), Xlibris Corporation, ISBN 1-479-7092-04

33. Tuller, David (1 November 2004) Dr. William Dobelle, Artificial Vision Pioneer, Dies at 62. New York Times

34. nurun.com. "Mr. Jen Naumann's high-tech paradise lost". Thewhig.com. Retrieved 19 December 2016.

35. Kennedy, PR; Bakay, RA (1998). "Restoration of neural output from a paralyzed patient by a direct brain connection". NeuroReport. 9 (8): 1707–11. doi:10.1097/00001756-199806010-00007. PMID 9665587.

36. Leigh R. Hochberg; Mijail D. Serruya; Friehs; Mukand; Saleh; Caplan; Branner; Chen; Penn; Donoghue (13 July 2006). Gerhard M. Friehs, Jon A. Mukand, Maryam Saleh, Abraham H. Caplan, Almut Branner, David Chen, Richard D. Penn and John P. Donoghue. "Neuronal ensemble control of prosthetic devices by a human with tetraplegia". Nature. 442 (7099): 164–171. Bibcode:2006Natur.442..164H. doi:10.1038/nature04970. PMID 16838014.

37. Hochberg, L. R.; Bacher, D.; Jarosiewicz, B.; Masse, N. Y.; Simeral, J. D.; Vogel, J.; Haddadin, S.; Liu, J.; Cash, S. S.; Van Der Smagt, P.; Donoghue, J. P. (2012). "Reach and grasp by people with tetraplegia using a neurally controlled robotic arm". Nature. 485 (7398): 372–5. Bibcode:2012Natur.485..372H. doi:10.1038/nature11076. PMC 3640850 . PMID 22596161.

38. Collinger, Jennifer L.; et al. (2013). "High-performance neuroprosthetic control by an individual with tetraplegia". The Lancet. 381 (9866): 557–564. doi:10.1016/S0140-6736(12)61816-9.

39. Gulati, Tanuj; Won, Seok Joon; Ramanathan, Dhakshin S.; Wong, Chelsea C.; Bodepudi, Anitha; Swanson, Raymond A.; Ganguly, Karunesh (2015). "Robust Neuroprosthetic Control from the Stroke Perilesional Cortex". The Journal of Neuroscience. 35 (22): 8653–8661. doi:10.1523/JNEUROSCI.5007-14.2015. PMID 26041930.

40. Serruya MD, Donoghue JP. (2003) Chapter III: Design Principles of a Neuromotor Prosthetic Device in Neuroprosthetics: Theory and Practice, ed. Kenneth W. Horch, Gurpreet S. Dhillon. Imperial College Press.

41. Teenager moves video icons just by imagination, press release, Washington University in St Louis, 9 October 2006

42. Schalk, G; Miller, KJ; Anderson, NR; Wilson, JA; Smyth, MD; Ojemann, JG; Moran, DW; Wolpaw, JR; Leuthardt, EC (2008). "Two-dimensional movement control using electrocorticographic signals in humans". Journal of Neural

Engineering.

43. Yanagisawa, Takafumi (2011). "Electrocorticograpic Control of Prosthetic Arm in Paralyzed Patients". American Neurological Association. doi:10.1002/ana.22613. ECoG- Based BCI has advantage in signal and durability that are absolutely necessary for clinical application
44. Pei, X. (2011). "Decoding Vowels and Consonants in Spoken and Imagined Words Using Electrocorticographic Signals in Humans". J Neural Eng 046028th ser. 8.4. PMID 21750369. Justin Williams, a biomedical engineer at the university, has already transformed the ECoG implant into a micro device that can be installed with a minimum of fuss. It has been tested in animals for a long period of time – the micro ECoG stays in place and doesn't seem to negatively affect the immune system.
45. Gallegos-Ayala, G; Furdea, A; Takano, K; Ruf, CA; Flor, H; Birbaumer, N (27 May 2014). "Brain communication in a completely locked-in patient using bedside near-infrared spectroscopy". Neurology. 82 (21): 1930–2. doi:10.1212/WNL.0000000000000449. PMID 24789862.
46. Mathôt, Sebastiaan; Melmi, Jean-Baptiste; Van Der Linden, Lotje; Van Der Stigchel, Stefan (2016). "The Mind-Writing Pupil: A Human-Computer Interface Based on Decoding of Covert Attention through Pupillometry". PLoS ONE. 11 (2): e0148805. doi:10.1371/journal.pone.0148805. PMID 26848745.
47. Just short of telepathy: can you interact with the outside world if you can't even blink an eye?, Psychology Today, May–June 2003
48. Yuan, H; Liu, Tao; Szarkowski, Rebecca; Rios, Cristina; Ashe, James; He, Bin (2010). "Negative covariation between task-related responses in alpha/beta-band activity and BOLD in human sensorimotor cortex: an EEG and fMRI study of motor imagery and movements". NeuroImage. 49 (3): 2596–2606. doi:10.1016/j.neuroimage.2009.10.028. PMC 2818527 . PMID 19850134.
49. Doud, AJ; Lucas, John P.; Pisansky, Marc T.; He, Bin (2011). Gribble, Paul L, ed. "Continuous Three-Dimensional Control of a Virtual Helicopter Using a Motor Imagery Based Brain-Computer Interface". PLoS ONE. 6 (10): e26322. Bibcode:2011PLoSO...626322D. doi:10.1371/journal.pone.0026322. PMC 3202533 . PMID 22046274.
50. "Thought-guided helicopter takes off". bbc.co.uk. 5 June 2013. Retrieved 5 June 2013.
51. Qin, L; Ding, Lei; He, Bin (2004). "Motor imagery classification by means of source analysis for brain-computer interface applications". Journal of Neural Engineering. 1 (3): 135–141. Bibcode:2004JNEng...1..135Q. doi:10.1088/1741-2560/1/3/002. PMID 15876632.
52. Höhne, J; Holz, E; Staiger-Sälzer, P; Müller, KR; Kübler, A; Tangermann, M (2014). "Motor imagery for severely motor-impaired patients: evidence for brain-computer interfacing as superior control solution.". PLOS ONE. 9 (8): e104854. Bibcode:2014PLoSO...9j4854H. doi:10.1371/journal.pone.0104854. PMC 4146550 . PMID 25162231.

53. Taheri, B; Knight, R; Smith, R (1994). "A dry electrode for EEG recording☆". Electroencephalography and Clinical Neurophysiology. 90 (5): 376–83. doi:10.1016/0013-4694(94)90053-1. PMID 7514984.

54. Alizadeh-Taheri, Babak (1994). "Active Micromachined Scalp Electrode Array for Eeg Signal Recording". PhD thesis. University of California: 82. Bibcode:1994PhDT........82A.

55. The Next BrainiacsWired Magazine, August 2001.

56. Pfurtscheller, G.; Müller, G. R.; Pfurtscheller, J. R.; Gerner, H. J. R.; Rupp, R. D. (2003). "'Thought' – control of functional electrical stimulation to restore hand grasp in a patient with tetraplegia". Neuroscience Letters. 351 (1): 33–36. doi:10.1016/S0304-3940(03)00947-9. PMID 14550907.

57. Do, An H; Wang, Po T; King, Christine E; Chun, Sophia N; Nenadic, Zoran (2013). "Brain-computer interface controlled robotic gait orthosis". Journal of NeuroEngineering and Rehabilitation. 10 (1): 111. doi:10.1186/1743-0003-10-111. ISSN 1743-0003.

58. Subject with Paraplegia Operates BCI-controlled RoGO (4x) at YouTube.com

59. Alex Blainey controls a cheap consumer robot arm using the EPOC headset via a serial relay port at YouTube.com

60. Artificial Neural Net Based Signal Processing for Interaction with Peripheral Nervous System Archived 30 September 2007 at the Wayback Machine.. In: Proceedings of the 1st International IEEE EMBS Conference on Neural Engineering. pp. 134–137. 20–22 March 2003.

61. Mental ways to make music, Cane, Alan, Financial Times, London (UK), 22 April 2005, p. 12

62. Ravi, K. V. R.; Palaniappan, R. (2005). "Neural network classification of late gamma band electroencephalogram features". Soft Computing. 10 (2): 163–169. doi:10.1007/s00500-004-0439-7.

63. New research to find out if your thoughts can be used to verify passwords Archived 21 December 2013 at the Wayback Machine.. Retrieved on 20 September 2012.

64. When mind over matter has a whole new meaning (From Gazette) Archived 30 May 2011 at the Wayback Machine.. Gazette-news.co.uk (13 April 2011). Retrieved on 29 May 2012.

65. Gürkök H.; Nijholt A. (2012). "Brain-Computer Interfaces for Multimodal Interaction: A Survey and Principles". Int. J. Hum. Comput. Interaction. 28 (5): 292–307. doi:10.1080/10447318.2011.582022.

66. Plass-Oude Bos, Danny; Reuderink, Boris; Van De Laar, Bram; Gürkök, Hayrettin; Mühl, Christian; Poel, Mannes; Nijholt, Anton; Heylen, Dirk (2010). "Brain-Computer Interfacing and Games". Brain-Computer Interfaces. Human-Computer Interaction Series. p. 149. doi:10.1007/978-1-84996-272-8_10. ISBN 978-1-84996-271-1.

67. Gürkök H., Nijholt A., Poel M. (2012). Brain-Computer Interface Games: Towards a Framework. ICEC. Lecture Notes in Computer Science. 2012. pp. 373–380.

doi:10.1007/978-3-642-33542-6_33. ISBN 978-3-642-33541-9.

68. Guger C., Ramoser H., Pfurtscheller G. (Dec 2000). "Real-time analysis with subject-specific spatial patterns". IEEE Trans Rehabil Eng. 8 (4): 447–56. doi:10.1109/86.895947. PMID 11204035.
69. Drummond, Katie (14 May 2009). "Pentagon Preps Soldier Telepathy Push". Wired Magazine. Retrieved 6 May 2009.
70. DARPA (May 2009). "Department of Defense Fiscal Year (FY) 2010 Budget Estimates May 2009". DARPA. Archived from the original on 30 August 2011. Retrieved 25 July 2011.
71. "The OpenEEG Project". Openeeg.sourceforge.net. Retrieved 19 December 2016.
72. "How To Hack Toy EEGs". Frontiernerds.com. Retrieved 19 December 2016.
73. Ranganatha Sitaram, Andrea Caria, Ralf Veit, Tilman Gaber, Giuseppina Rota, Andrea Kuebler and Niels Birbaumer(2007) "FMRI Brain–Computer Interface: A Tool for Neuroscientific Research and Treatment"
74. Peplow, Mark (2004). "Mental ping-pong could aid paraplegics". News@nature. doi:10.1038/news040823-18.
75. To operate robot only with brain, ATR and Honda develop BMI base technology, Tech-on, 26 May 2006
76. Miyawaki, Yoichi; Uchida, Hajime; Yamashita, Okito; Sato, Masa-aki; Morito, Yusuke; Tanabe, Hiroki C.; Sadato, Norihiro; Kamitani, Yukiyasu (2008). "Visual Image Reconstruction from Human Brain Activity using a Combination of Multiscale Local Image Decoders". Neuron. 60 (5): 915. doi:10.1016/j.neuron.2008.11.004. PMID 19081384.
77. Nishimoto, Shinji; Vu, An T.; Naselaris, Thomas; Benjamini, Yuval; Yu, Bin; Gallant, Jack L. (2011). "Reconstructing Visual Experiences from Brain Activity Evoked by Natural Movies". Current Biology. 21 (19): 1641. doi:10.1016/j.cub.2011.08.031. PMID 21945275.
78. Yam, Philip (22 September 2011). "Breakthrough Could Enable Others to Watch Your Dreams and Memories". Scientific American. Retrieved 25 September 2011.
79. "Reconstructing visual experiences from brain activity evoked by natural movies (Project page)". The Gallant Lab at UC Berkeley. Retrieved 25 September 2011.
80. Yasmin Anwar (22 September 2011). "Scientists use brain imaging to reveal the movies in our mind". UC Berkeley News Center. Retrieved 25 September 2011.
81. Video on YouTube
82. "Neurogaming".
83. Video on YouTube
84. "Merging Cognitive Neuroscience & Virtual Simulation in an Interactive Training Platform".
85. "Neurogamingconf.Com". Neurogamingconf.Com. Retrieved 19 December 2016.
86. Coyle, Damien; Marshall, David; Wilson, Shane; Callaghan, Michael. "Games, Gameplay, and BCI: The State of the Art". 5 (2): 83. doi:10.1109/

TCIAIG.2013.2263555.

87. Ang, Kai Keng; Chin, Zheng Yang; Wang, Chuanchu; Guan, Cuntai; Zhang, Haihong (1 January 2012). "Filter bank common spatial pattern algorithm on BCI competition IV Datasets 2a and 2b". Neuroprosthetics. 6: 39. doi:10.3389/fnins.2012.00039. PMC 3314883 . PMID 22479236.
88. Kennedy, Pagan (18 September 2011). "The Cyborg in Us All". New York Times. Retrieved 28 January 2012.
89. Bland, Eric (13 October 2008). "Army Developing'synthetic telepathy'". Discovery News. Retrieved 13 October 2008.
90. Pais-Vieira, Miguel; Lebedev, Mikhail; Kunicki, Carolina; Wang, Jing; Nicolelis, Miguel A.L. (28 February 2013). "A Brain-to-Brain Interface for Real-Time Sharing of Sensorimotor Information". Scientific Reports. 3. Bibcode:2013NatSR...3E1319P. doi:10.1038/srep01319.
91. Gorman, James (28 February 2013). "One Rat Thinks, and Another Reacts". New York Times. Retrieved 28 February 2013.
92. "Brain-to-brain interface lets rats share information via internet". The Guardian. 1 March 2013. Retrieved 2 March 2013.
93. "Direct brain-to-brain communication demonstrated in human subjects". Science Daily. 3 September 2014. Retrieved 9 September 2014.
94. Grau, Carles; Ginhoux, Romuald; Riera, Alejandro; Nguyen, Thanh Lam; Chauvat, Hubert; Berg, Michel; Amengual, Julià L.; Pascual-Leone, Alvaro; Ruffini, Giulio (19 August 2014). "Conscious Brain-to-Brain Communication in Humans Using Non-Invasive Technologies". PLoS. 9 (8): e105225. Bibcode:2014PLoSO...9j5225G. doi:10.1371/journal.pone.0105225.
95. Tressoldi, Patrizio E.; Pederzoli, Luciano; Bilucaglia, Marco; Caini, Patrizio; Fedele, Pasquale; Ferrini, Alessandro; Melloni, Simone; Accardo, Agostino Patrizio (24 May 2014). "Brain-to-Brain (Mind-to-Mind) Interaction at Distance: A Pilot Study". Rochester, NY: Social Science Research Network.
96. Tressoldi, Patrizio E.; Pederzoli, Luciano; Bilucaglia, Marco; Caini, Patrizio; Fedele, Pasquale; Ferrini, Alessandro; Melloni, Simone; Richeldi, Diana; Richeldi, Florentina (7 March 2014). "Brain-to-Brain (Mind-to-Mind) Interaction at Distance: A Confirmatory Study". Rochester, NY: Social Science Research Network.
97. Mazzatenta, A.; Giugliano, M.; Campidelli, S.; Gambazzi, L.; Businaro, L.; Markram, H.; Prato, M.; Ballerini, L. (2007). "Interfacing Neurons with Carbon Nanotubes: Electrical Signal Transfer and Synaptic Stimulation in Cultured Brain Circuits". Journal of Neuroscience. 27 (26): 6931–6. doi:10.1523/JNEUROSCI.1051-07.2007. PMID 17596441.
98. Press release, Caltech, 27 October 1997
99. Coming to a brain near you Archived 10 September 2006 at the Wayback Machine., Wired News, 22 October 2004
100. 'Brain' in a dish flies flight simulator, CNN, 4 November 2004

101. Clausen, Jens (2009). "Man, machine and in between". Nature. 457 (7233): 1080–1081. Bibcode:2009Natur.457.1080C. doi:10.1038/4571080a.

102. Haselager, Pim; Vlek, Rutger; Hill, Jeremy; Nijboer, Femke (2009). "A note on ethical aspects of BCI". Neural Networks. 22 (9): 1352–1357. doi:10.1016/j.neunet.2009.06.046.

103. Tamburrini, Guglielmo (2009). "Brain to Computer Communication: Ethical Perspectives on Interaction Models". Neuroethics. 2 (3): 137–149. doi:10.1007/s12152-009-9040-1.

104. Nijboer, Femke; Clausen, Jens; Allison, Brendan Z; Haselager, Pim (2011). "Stakeholders' opinions on ethical issues related to brain-computer interfacing". Neuroethics. 6 (3): 541–578. doi:10.1007/s12152-011-9132-6.

105. Ang, Kai Keng; Guan, Cuntai; Chua, Karen Sui Geok; Ang, Beng Ti; Kuah, Christopher Wee Keong; Wang, Chuanchu; Phua, Kok Soon; Chin, Zheng Yang; Zhang, Haihong (1 October 2011). "A Large Clinical Study on the Ability of Stroke Patients to Use an EEG-Based Motor Imagery Brain-Computer Interface". Clinical EEG and Neuroscience. 42 (4): 253–258. doi:10.1177/155005941104200411. PMID 22208123.

106. ""intendiX-SOCI": Introduces Mind-controlled Computer Gaming at CeBIT2012". PR Newswire. 5 March 2012.

107. Fedele, Pasquale; Fedele, Chiara; Fath, Jarrod (2014). "Braincontrol Basic Communicator: A Brain-Computer Interface Based Communicator for People with Severe Disabilities". Universal Access in Human-Computer Interaction. Design and Development Methods for Universal Access. Lecture Notes in Computer Science. 8513. pp. 487–494. doi:10.1007/978-3-319-07437-5_46. ISBN 978-3-319-07436-8.

108. Fedele, P. et al. (2016) "Results of a 3 year study of a BCI-based communicator for patients with severe disabilities", ACHI 2016, The Ninth International Conference on Advances in Computer-Human Interactions, IARIA Journal

109. Fedele, P. and Gioia, M. (2016) "Results of a 3 years study of a BCI-based communicator for advanced ALS patients". Proceedings of the International Society for Augmentative and Alternative Communication (ISAAC)

110. "Sony patent neural interface". Archived from the original on 7 April 2012.

111. "Mind Games". The Economist. 23 March 2007.

112. "nia Game Controller Product Page". OCZ Technology Group. Retrieved 30 January 2013.

113. Li, Shan (8 August 2010). "Mind reading is on the market". Los Angeles Times.

114. Brains-on with NeuroSky and Square Enix's Judecca mind-control game. Engadget.com (9 October 2008). Retrieved on 29 May 2012.

115. New games powered by brain waves. Physorg.com (10 January 2009). Retrieved on 2010-09-12.

116. Snider, Mike (7 January 2009). "Toy trains 'Star Wars' fans to use The Force". USA Today. Retrieved 1 May 2010.

117. "Emotiv Homepage". Emotiv.com. Retrieved 29 December 2009.
118. "necomimi" selected "TIME MAGAZINE / The 50 best invention of the year". Neurowear.com. Retrieved on 29 May 2012.
119. "LIFESUIT Updates & News – They Shall Walk". Theyshallwalk.org. Retrieved 19 December 2016.
120. "BNCI Horizon 2020: towards a roadmap for the BCI community". Brain-Computer Interfaces. 2: 1–10. doi:10.1080/2326263X.2015.1008956.
121. Allison, B.Z., Dunne, S., Leeb, R., Millan, J., and Nijholt, A. (2013). Towards Practical Brain-Computer Interfaces: Bridging the Gap from Research to Real-World Applications. Springer Verlag, Berlin Heidelberg. ISBN 978-3-642-29746-5.
122. Guger, C., Allison, B.Z., and Edlinger, G. (2013). Brain-Computer Interface Research: A State-of-the-Art Summary. Springer Verlag, Berlin Heidelberg.
123. Guger, C., Allison, B.Z., Leuthardt, E.C., and Edlinger, G. (2014). The BCI Award 2012: A State-of-the-Art Summary 2. Springer Verlag, Berlin Heidelberg.
124. Guger, C., Allison, B.Z., and Vaughan, T.M. (2014). The BCI Award 2013: A State-of-the-Art Summary 3. Springer Verlag, Berlin Heidelberg.
125. Edlinger, G., Allison, B.Z., and Guger, C. (2015). "How many people could use a BCI system?" pp. 33–66 in Clinical Systems Neuroscience, Kansaku, K., Cohen, L., and Birbaumer, N. (eds.) Springer Verlag Japan: Tokyo. ISBN 978-4-431-55037-2.
126. Chatelle, Camille; Chennu, Srivas; Noirhomme, Quentin; Cruse, Damian; Owen, Adrian M.; Laureys, Steven (2012). "Brain–computer interfacing in disorders of consciousness". Brain Injury. 26 (12): 1510. doi:10.3109/02699052.2012.698362. PMID 22759199.
127. Boly M, Massimini M, Garrido MI, Gosseries O, Noirhomme Q, Laureys S, Soddu A (2012). "Brain connectivity in disorders of consciousness". Brain Connectivity. 2 (1): 1to10. doi:10.1089/brain.2011.0049.
128. Gibson, Raechelle M.; Fernã¡Ndez-Espejo, Davinia; Gonzalez-Lara, Laura E.; Kwan, Benjamin Y.; Lee, Donald H.; Owen, Adrian M.; Cruse, Damian (2014). "Multiple tasks and neuroimaging modalities increase the likelihood of detecting covert awareness in patients with disorders of consciousness". Frontiers in Human Neuroscience. 8. doi:10.3389/fnhum.2014.00950.
129. Risetti, Monica; Formisano, Rita; Toppi, Jlenia; Quitadamo, Lucia R.; Bianchi, Luigi; Astolfi, Laura; Cincotti, Febo; Mattia, Donatella (2013). "On ERPs detection in disorders of consciousness rehabilitation". Frontiers in Human Neuroscience. 7. doi:10.3389/fnhum.2013.00775.
130. Radzik, Iwona; Miziak, Barbara; Dudka, Jarosław; Chrościńska-Krawczyk, Magdalena; Czuczwar, Stanisław J. (2015). "Prospects of epileptogenesis prevention". Pharmacological Reports. 67 (3): 663. doi:10.1016/j.pharep.2015.01.016. PMID 25933984.
131. Ritaccio, Anthony; Brunner, Peter; Gunduz, Aysegul; Hermes, Dora; Hirsch, Lawrence J.; Jacobs, Joshua; Kamada, Kyousuke; Kastner, Sabine; Knight,

Robert T.; Lesser, Ronald P.; Miller, Kai; Sejnowski, Terrence; Worrell, Gregory; Schalk, Gerwin (2014). "Proceedings of the Fifth International Workshop on Advances in Electrocorticography". Epilepsy & Behavior. 41: 183. doi:10.1016/j.yebeh.2014.09.015.

132. What is computational neuroscience? Patricia S. Churchland, Christof Koch, Terrence J. Sejnowski. in Computational Neuroscience pp.46-55. Edited by Eric L. Schwartz. 1993. MIT Press
133. Schwartz, Eric (1990). Computational neuroscience. Cambridge, Mass: MIT Press. ISBN 0-262-19291-8.
134. Bower, James M. (2013). 20 years of Computational neuroscience. Berlin, Germany: Springer. ISBN 978-1461414230.
135. Lapicque L (1907). "Recherches quantitatives sur l'excitation électrique des nerfs traitée comme une polarisation". J. Physiol. Pathol. Gen. 9: 620–635.
136. Brunel N, Van Rossum MC (2007). "Lapicque's 1907 paper: from frogs to integrate-and-fire". Biol. Cybern. 97 (5–6): 337–339. doi:10.1007/s00422-007-0190-0. PMID 17968583.
137. Hubel DH, Wiesel TN (1962). "Receptive fields, binocular interaction and functional architecture in the cat's visual cortex". J. Physiol. (Lond.). 160 (1): 106–54. doi:10.1113/jphysiol.1962.sp006837. PMC 1359523 . PMID 14449617.
138. Forrest MD (2014). "Intracellular Calcium Dynamics Permit a Purkinje Neuron Model to Perform Toggle and Gain Computations Upon its Inputs.". Frontiers in Computational Neuroscience. 8: 86. doi:10.3389/fncom.2014.00086. PMC 4138505 . PMID 25191262.
139. Wu, Samuel Miao-sin; Johnston, Daniel (1995). Foundations of cellular neurophysiology. Cambridge, Mass: MIT Press. ISBN 0-262-10053-3.
140. Koch, Christof (1999). Biophysics of computation: information processing in single neurons. Oxford: Oxford University Press. ISBN 0-19-510491-9.
141. Forrest MD (2014). "Intracellular Calcium Dynamics Permit a Purkinje Neuron Model to Perform Toggle and Gain Computations Upon its Inputs.". Frontiers in Computational Neuroscience. 8: 86. doi:10.3389/fncom.2014.00086. PMC 4138505 . PMID 25191262.
142. Forrest MD (April 2015). "Simulation of alcohol action upon a detailed Purkinje neuron model and a simpler surrogate model that runs >400 times faster". BMC Neuroscience. 16 (27). doi:10.1186/s12868-015-0162-6.
143. Chklovskii DB, Mel BW, Svoboda K (October 2004). "Cortical rewiring and information storage". Nature. 431 (7010): 782–8. Bibcode:2004Natur.431..782C. doi:10.1038/nature03012. PMID 15483599.
144. Review article
145. Durstewitz D, Seamans JK, Sejnowski TJ (2000). "Neurocomputational models of working memory". Nat Neurosci. 3 (Suppl): 1184–91. doi:10.1038/81460. PMID 11127836.
146. Fusi S, Drew PJ, Abbott LF (2005). "Cascade models of synaptically stored

memories". Neuron. 45 (4): 599–611. doi:10.1016/j.neuron.2005.02.001. PMID 15721245.

147. Coggan JS, Bartol TM, Esquenazi E, et al. (2005). "Evidence for ectopic neurotransmission at a neuronal synapse". Science. 309 (5733): 446–51. Bibcode:2005Sci...309..446C. doi:10.1126/science.1108239. PMC 2915764 . PMID 16020730.

148. Schneidman E, Berry MJ, Segev R, Bialek W (2006). "Weak pairwise correlations imply strongly correlated network states in a neural population". Nature. 440 (7087): 1007–12. arXiv:q-bio/0512013 . Bibcode:2006Natur.440.1007S. doi:10.1038/nature04701. PMC 1785327 . PMID 16625187.

149. Anderson, Charles H.; Eliasmith, Chris (2004). Neural Engineering: Computation, Representation, and Dynamics in Neurobiological Systems (Computational Neuroscience). Cambridge, Mass: The MIT Press. ISBN 0-262-55060-1.

150. Machens CK, Romo R, Brody CD (2005). "Flexible control of mutual inhibition: a neural model of two-interval discrimination". Science. 307 (5712): 1121–4. Bibcode:2005Sci...307.1121M. doi:10.1126/science.1104171. PMID 15718474.

151. Robinson PA, Rennie CJ, Rowe DL, O'Connor SC, Gordon E (2005). "Multiscale brain modelling". Philosophical Transactions of the Royal Society B. 360 (1457): 1043–1050. doi:10.1098/rstb.2005.1638. PMC 1854922 . PMID 16087447.

152. Crick F, Koch C (2003). "A framework for consciousness". Nat Neurosci. 6 (2): 119–26. doi:10.1038/nn0203-119. PMID 12555104.

153. Adaszewski S1, Dukart J, Kherif F, Frackowiak R, Draganski B; Alzheimer's Disease Neuroimaging Initiative (2013). "How early can we predict Alzheimer's disease using computational anatomy?". Neurobiol Aging. 34 (12): 2815–26. doi:10.1016/j.neurobiolaging.2013.06.015.

154. Friston KJ, Stephan KE, Montague R, Dolan RJ (2014). "Computational psychiatry: the brain as a phantastic organ". Lancet Psych. 1 (2): 148–58. doi:10.1016/S2215-0366(14)70275-5.

155. Dupuis, Paul; Grenander, Ulf; Miller, Michael. "Variational Problems on Flows of Diffeomorphisms for Image Matching". ResearchGate. Retrieved 2016-02-20.

156. Arnold, V. (1966). "Sur la géomérie différentielle des groupes de Lie de dimension infinie et ses applications à l'hydrodynamique des fluides parfaits". Ann. Inst. Fourier (in French). 16 (1): 319–361. doi:10.5802/aif.233. MR 202082.

157. Laurent Younes. Shapes and Diffeomorphisms. Springer. ISBN 9783642120541.

158. Durer, Albrecht (1528-01-01). Hierinn sind begriffen vier Bucher von menschlicher Proportion durch Albrechten Durer von Nurerberg erfunden und beschuben zu nutz allen denen so zu diser kunst lieb tragen. Hieronymus Andreae Formschneider. Check date values in: |date= (help)

159. Library, State of Texas, University of Texas Health Science Center at San Antonio. "Albrecht Dürer's Human Proportions « UT Health Science Center Library". library.uthscsa.edu. Retrieved 2016-03-16.

160. "Albrecht Dürer". The Morgan Library & Museum. Retrieved 2016-03-16.

161. Grenander, Ulf; Miller, Michael I. (1998-12-01). "Computational Anatomy: An Emerging Discipline". Q. Appl. Math. 56 (4): 617–694.

162. "Brown University -- Pattern Theory Group: Home". www.dam.brown.edu. Retrieved 2015-12-27.

163. Miller, Michael I.; Younes, Laurent; Trouvé, Alain (2014-03-01). "Diffeomorphometry and geodesic positioning systems for human anatomy". Technology. 2 (1): 36–43. doi:10.1142/S2339547814500010. PMC 4041578 . PMID 24904924.

164. Grenander, Ulf. General Pattern Theory : A Mathematical Study of Regular Structures. Oxford University Press. ISBN 9780198536710.

165. U. Grenander and M. I. Miller (2007-02-08). Pattern Theory: From Representation to Inference. Oxford: Oxford University Press. ISBN 9780199297061.

166. Van Essen, D. C.; Maunsell, J. H. (1980-05-15). "Two-dimensional maps of the cerebral cortex". The Journal of Comparative Neurology. 191 (2): 255–281. doi:10.1002/cne.901910208. PMID 7410593.

167. Bajcsy, Ruzena; Kovačič, Stane (1989-04-01). "Multiresolution Elastic Matching". Comput. Vision Graph. Image Process. 46 (1): 1–21. doi:10.1016/S0734-189X(89)80014-3.

168. Bookstein, F. L. (1989-06-01). "Principal Warps: Thin-Plate Splines and the Decomposition of Deformations". IEEE Trans. Pattern Anal. Mach. Intell. 11 (6): 567–585. doi:10.1109/34.24792.

169. Christensen, Gary; Rabbitt, Richard; Miller, Michael I. (1993-01-01). Jerry Prince, ed. A deformable neuroanatomy textbook based on viscous fluid mechanics: Proceedings of the ... Conference on Information Sciences and Systems. Department of Electrical Engineering, Johns Hopkins University.

170. Christensen, G. E.; Rabbitt, R. D.; Miller, M. I. (1996-10-01). "Deformable Templates Using Large Deformation Kinematics". Trans. Img. Proc. 5 (10): 1435–1447. doi:10.1109/83.536892. PMID 18290061.

171. Miller, Michael; Joshi, Sarang; Christensen; Brain Warping Book Author: Toga, Arthur (1997). Brain Warping: Chapter 7: Large Deformation Fluid Diffeomorphisms for Landmark and Image Matching. p. 115.

172. Walter Freiberger (ed.). "Current and Future Challenges in the Applications of Mathematics". Quarterly of Applied Mathematics.

173. "Colloque Mathematiques et reconnaissance de formes". www.ceremade.dauphine.fr. Retrieved 2015-12-19.

174. "Robert Azencott, mathématicien polyglotte | La Recherche". www.larecherche.fr. Retrieved 2016-02-20.

175. Trouve, Alain. "An Approach of Pattern Recognition through Infinite Dimensional Diffeomorphism" (PDF).

176. Dupuis, Paul; Grenander, Ulf (1998-09-01). "Variational Problems on Flows of Diffeomorphisms for Image Matching". Q. Appl. Math. LVI (3): 587–600.

177. Miller, Michael I.; Trouve, Alain; Younes, Laurent (2002-01-01). "On the

metrics and euler-lagrange equations of computational anatomy". Annual Review of Biomedical Engineering. 4: 375–405. doi:10.1146/annurev.bioeng.4.092101.125733. PMID 12117763.

178. Miller, Michael I.; Trouvé, Alain; Younes, Laurent (2006-01-28). "Geodesic Shooting for Computational Anatomy". Journal of Mathematical Imaging and Vision. 24 (2): 209–228. doi:10.1007/s10851-005-3624-0. PMC 2897162 . PMID 20613972.
179. Miller, Michael I.; Trouvé, Alain; Younes, Laurent (2015-12-07). "Hamiltonian Systems and Optimal Control in Computational Anatomy: 100 Years Since D'Arcy Thompson". Annual Review of Biomedical Engineering. 17: 447–509. doi:10.1146/annurev-bioeng-071114-040601. PMID 26643025.
180. Miller, M. I.; Younes, L. (2001-01-01). "Group Actions, Homeomorphisms, and Matching: A General Framework". Int. J. Comput. Vision. 41 (1-2): 61–84. doi:10.1023/A:1011161132514.
181. Trouvé, A.; Younes, L. (2005-01-01). "Local Geometry of Deformable Templates". SIAM Journal on Mathematical Analysis. 37 (1): 17–59. doi:10.1137/S0036141002404838.
182. Micheli, Mario; Michor, Peter W.; Mumford, David (2012-03-01). "Sectional Curvature in Terms of the Cometric, with Applications to the Riemannian Manifolds of Landmarks". SIAM J. Img. Sci. 5 (1): 394–433. doi:10.1137/10081678X.
183. "Home Page". cnls.lanl.gov. Retrieved 2015-12-19.
184. Joshi, S. C.; Miller, M. I. (2000-01-01). "Landmark matching via large deformation diffeomorphisms". IEEE transactions on image processing: a publication of the IEEE Signal Processing Society. 9 (8): 1357–1370. doi:10.1109/83.855431. PMID 18262973.
185. Holm, Darryl D. (2009-08-29). "Peakons". arXiv:0908.4351 .
186. Ebin, David G.; Marsden, Jerrold E. (1969-09-01). "Groups of diffeomorphisms and the solution of the classical Euler equations for a perfect fluid". Bulletin of the American Mathematical Society. 75 (5): 962–967. doi:10.1090/s0002-9904-1969-12315-3.
187. Mumford, David; Michor, Peter W. (2012-09-28). "On Euler's equation and 'EPDiff'". arXiv:1209.6576 . doi:10.3934/jgm.2013.5.xx.
188. Scherzer, Otmar (2010-11-23). Handbook of Mathematical Methods in Imaging. Springer Science & Business Media. ISBN 9780387929194.
189. Glaunes, J.; Trouve, A.; Younes, L. (2004-06-01). "Diffeomorphic matching of distributions: a new approach for unlabelled point-sets and sub-manifolds matching". Proceedings of the 2004 IEEE Computer Society Conference on Computer Vision and Pattern Recognition, 2004. CVPR 2004. 2: II–712–II–718 Vol.2. doi:10.1109/CVPR.2004.1315234.
190. "Landmark Matching via Large Deformation Diffeomorphisms on the Sphere: Special Issue on Mathematics and Image Analysis". ResearchGate. doi:10.1023/B:JMIV.0000011326.88682.e5. Retrieved 2016-03-27.

191. Du, Jia; Younes, Laurent; Qiu, Anqi (2011-05-01). "Whole brain diffeomorphic metric mapping via integration of sulcal and gyral curves, cortical surfaces, and images". NeuroImage. 56 (1): 162–173. doi:10.1016/j.neuroimage.2011.01.067. PMC 3119076 . PMID 21281722.

192. Vaillant, Marc; Glaunès, Joan (2005-01-01). "Surface matching via currents". Information Processing in Medical Imaging: Proceedings of the ... Conference. 19: 381–392. PMID 17354711.

193. Vaillant, Marc; Qiu, Anqi; Glaunès, Joan; Miller, Michael I. (2007-02-01). "Diffeomorphic Metric Surface Mapping in Superior Temporal Gyrus". NeuroImage. 34 (3): 1149–1159. doi:10.1016/j.neuroimage.2006.08.053. PMC 3140704 . PMID 17185000.

194. Durrleman, Stanley; Pennec, Xavier; Trouvé, Alain; Ayache, Nicholas (2009-10-01). "Statistical models of sets of curves and surfaces based on currents". Medical Image Analysis. 13 (5): 793–808. doi:10.1016/j.media.2009.07.007. PMID 19679507.

195. M.F. Beg and M. I. Miller and A. Trouve and L. Younes. "Computing Large Deformation Metric Mappings via Geodesic Flows of Diffeomorphisms". ResearchGate. doi:10.1023/B:VISI.0000043755.93987.aa. Retrieved 2016-01-27.

196. Cao, Yan; Miller, Michael I.; Mori, Susumu; Winslow, Raimond L.; Younes, Laurent (2006-07-05). "Diffeomorphic Matching of Diffusion Tensor Images". Proceedings / CVPR, IEEE Computer Society Conference on Computer Vision and Pattern Recognition. IEEE Computer Society Conference on Computer Vision and Pattern Recognition. 2006: 67. doi:10.1109/CVPRW.2006.65. PMC 2920614 . PMID 20711423.

197. Charon, Nicolas; Trouvé, Alain. "The varifold representation of non-oriented shapes for diffeomorphic registration". SIAM Journal on Imaging Sciences. 6 (4): 2547–2580. arXiv:1304.6108 . doi:10.1137/130918885.

198. Miller, Michael I. (2004-01-01). "Computational anatomy: shape, growth, and atrophy comparison via diffeomorphisms". NeuroImage. 23 Suppl 1: S19–33. doi:10.1016/j.neuroimage.2004.07.021. PMID 15501089.

199. Trouvé, Alain; Vialard, François-Xavier (2010-03-19). "Shape Splines and Stochastic Shape Evolutions: A Second Order Point of View". arXiv:1003.3895 .

200. Fletcher, P.T.; Lu, C.; Pizer, S.M.; Joshi, S. (2004-08-01). "Principal geodesic analysis for the study of nonlinear statistics of shape". IEEE Transactions on Medical Imaging. 23 (8): 995–1005. doi:10.1109/TMI.2004.831793.

201. "Ueber die Hypothesen, welche der Geometrie zu Grunde liegen.". www.maths.tcd.ie. Retrieved 2016-03-16.

202. Bernhard Riemann. "Über die Hypothesen, welche der Geometrie". Springer. ISBN 9783642351204.

203. Peter W. Michor (2008-07-23). Topics in Differential Geometry. American Mathematical Society. ISBN 9780821820032.

204. NeuroImage. 23 (Supplement 1): S1–S300. 2004 http://www.sciencedirect.com/

science/journal/10538119/23/supp/S1. Missing or

205. NeuroImage. 45 (Supplement 1): S1–S222. 2009 http://www.sciencedirect.com/science/journal/10538119/45/1/supp/S1. Missing or

206. Fonseca, Carissa G.; Backhaus, Michael; Bluemke, David A.; Britten, Randall D.; Chung, Jae Do; Cowan, Brett R.; Dinov, Ivo D.; Finn, J. Paul; Hunter, Peter J. (2011-08-15). "The Cardiac Atlas Project--an imaging database for computational modeling and statistical atlases of the heart". Bioinformatics (Oxford, England). 27 (16): 2288–2295. doi:10.1093/bioinformatics/btr360. PMC 3150036 . PMID 21737439.

207. "CellOrganizer 1.8 Release Notes" (PDF).

208. Jamie Weir; et al. (2010-03-09). Imaging Atlas of Human Anatomy (4th ed.). Edinburgh: Mosby. ISBN 9780723434573.

209. "The Whole Brain Atlas". www.med.harvard.edu. Retrieved 2016-01-26.

210. Mazziotta, J; Toga, A; Evans, A; Fox, P; Lancaster, J; Zilles, K; Woods, R; Paus, T; Simpson, G (2001-08-29). "A probabilistic atlas and reference system for the human brain: International Consortium for Brain Mapping (ICBM).". Philosophical Transactions of the Royal Society of London. Series B. 356 (1412): 1293–1322. doi:10.1098/rstb.2001.0915. PMC 1088516 . PMID 11545704.

211. "White Matter Atlas - Diffusion Tensor Imaging Atlas of the Brain's White Matter Tracts". www.dtiatlas.org. Retrieved 2016-01-26.

212. Miller, M I; Christensen, G E; Amit, Y; Grenander, U (1993-12-15). "Mathematical textbook of deformable neuroanatomies.". Proceedings of the National Academy of Sciences of the United States of America. 90 (24): 11944–11948. doi:10.1073/pnas.90.24.11944. PMC 48101 . PMID 8265653.

213. "FreeSurfer". freesurfer.net. Retrieved 2015-12-08.

214. "FSL - FslWiki". fsl.fmrib.ox.ac.uk. Retrieved 2015-12-08.

215. "NITRC: MRI Studio: Tool/Resource Info". www.nitrc.org. Retrieved 2015-12-08.

216. "SPM software - Statistical Parametric Mapping". www.fil.ion.ucl.ac.uk. Retrieved 2015-12-08.

217. "stnava/ANTs". GitHub. Retrieved 2015-12-11.

218. Ashburner, John (2007-10-15). "A fast diffeomorphic image registration algorithm". NeuroImage. 38 (1): 95–113. doi:10.1016/j.neuroimage.2007.07.007. PMID 17761438.

219. "Software - Tom Vercauteren". sites.google.com. Retrieved 2015-12-11.

220. "NITRC: LDDMM: Tool/Resource Info". www.nitrc.org. Retrieved 2015-12-11.

221. "Publication:Comparing algorithms for diffeomorphic registration: Stationary LDDMM and Diffeomorphic Demons". www.openaire.eu. Retrieved 2015-12-11.

222. Zhang, Miaomiao; Fletcher, P. Thomas (2015-01-01). "Finite-Dimensional Lie Algebras for Fast Diffeomorphic Image Registration". Information Processing in Medical Imaging: Proceedings of the ... Conference. 24: 249–259. ISSN 1011-2499. PMID 26221678.

223. Christensen, G. E.; Rabbitt, R. D.; Miller, M. I. (1996-10-01). "Deformable Templates Using Large Deformation Kinematics". Trans. Img. Proc. 5 (10): 1435–1447. doi:10.1109/83.536892. PMID 18290061.

224. GE Christensen,RD Rabbitt,MI Miller, Deformable templates using large deformation kinematics, IEEE Trans Image Process. 1996;5(10):1435-47.

225. GE Christensen, SC Joshi, MI Miller, Volumetric transformation of brain anatomy IEEE Transactions on Medical Imaging,1997.

226. P. Dupuis, U. Grenander, M.I. Miller, Existence of Solutions on Flows of Diffeomorphisms, Quarterly of Applied Math, 1997.

227. A. Trouvé. Action de groupe de dimension infinie et reconnaissance de formes. C R Acad Sci Paris Sér I Math, 321(8):1031– 1034, 1995.

228. Younes, L. (1998-04-01). "Computable Elastic Distances Between Shapes". SIAM Journal on Applied Mathematics. 58 (2): 565–586. doi:10.1137/S0036139995287685.

229. Mio, Washington; Srivastava, Anuj; Joshi, Shantanu (2006-09-25). "On Shape of Plane Elastic Curves". International Journal of Computer Vision. 73 (3): 307–324. doi:10.1007/s11263-006-9968-0.

230. Michor, Peter W.; Mumford, David; Shah, Jayant; Younes, Laurent (2007-06-28). "A Metric on Shape Space with Explicit Geodesics". arXiv:0706.4299 .

231. Michor, Peter W.; Mumford, David. "An overview of the Riemannian metrics on spaces of curves using the Hamiltonian approach". Applied and Computational Harmonic Analysis. 23 (1): 74–113. arXiv:math/0605009 . doi:10.1016/j.acha.2006.07.004.

232. Kurtek, Sebastian; Klassen, Eric; Gore, John C.; Ding, Zhaohua; Srivastava, Anuj (2012-09-01). "Elastic geodesic paths in shape space of parameterized surfaces". IEEE transactions on pattern analysis and machine intelligence. 34 (9): 1717–1730. doi:10.1109/TPAMI.2011.233. PMID 22144521.

233. Miller, Michael I.; Trouvé, Alain; Younes, Laurent (2015-01-01). "Hamiltonian Systems and Optimal Control in Computational Anatomy: 100 Years Since D'arcy Thompson". Annual Review of Biomedical Engineering. 17 (1): null. doi:10.1146/annurev-bioeng-071114-040601. PMID 26643025.

234. Miller, M. I.; Younes, L. (2001-01-01). "Group Actions, Homeomorphisms, And Matching: A General Framework". International Journal of Computer Vision. 41: 61–84.

235. MILLER, MICHAEL I.; TROUVÉ, ALAIN; YOUNES, LAURENT (2006-01-31). "Geodesic Shooting for Computational Anatomy". Journal of mathematical imaging and vision. 24 (2): 209–228. doi:10.1007/s10851-005-3624-0. PMC 2897162 . PMID 20613972.

236. M.I. Miller, A. Trouve, L. Younes, Geodesic Shooting in Computational Anatomy, IJCV, 2006.

237. Holm DD, Marsden JE, Ratiu TS. 1998. The Euler-Poincaré equations and semidirect products with applications to continuum theories. Adv. Math. 137:1–81

238. Miller, M.I.; Trouve, A.; Younes, L (2002). "On the Metrics and Euler-Lagrange equations of Computational Anatomy". Annu. Rev. Biomed. Eng. 4: 375–405. doi:10.1146/annurev.bioeng.4.092101.125733. PMID 12117763.

239. Glaunès J, Trouvé A, Younes L. 2006. Modeling planar shape variation via Hamiltonian flows of curves. In Statistics and Analysis of Shapes, ed. H Krim, A Yezzi Jr, pp. 335–61. Model. Simul. Sci. Eng. Technol. Boston: Birkhauser

240. Arguillère S, Trélat E, Trouvé A, Younes L. 2014. Shape deformation analysis from the optimal control viewpoint. arXiv:1401.0661

241. Michael I. Miller, Laurent Younes, and Alain Trouvé, Diffeomorphometry and geodesic positioning systems for human anatomy,Technology 02, 36 (2014). DOI: 10.1142/S2339547814500010

242. Michor, Peter W.; Mumford, David (2007-07-01). "An overview of the Riemannian metrics on spaces of curves using the Hamiltonian approach". Applied and Computational Harmonic Analysis. Special Issue on Mathematical Imaging. 23 (1): 74–113. doi:10.1016/j.acha.2006.07.004.

243. S. Joshi and M.I. Miller, Landmark matching via large deformation diffeomorphisms, IEEE Trans Image Process. 2000;9(8):1357-70. doi: 10.1109/83.855431.

244. V. Camion, L. Younes: Geodesic Interpolating Splines (EMMCVPR 2001)

245. J Glaunès, M Vaillant, MI Miller. Landmark matching via large deformation diffeomorphisms on the sphere Journal of mathematical imaging and vision, 2004.

246. Miller, Michael I.; Trouvé, Alain; Younes, Laurent (2015-01-01). "Hamiltonian Systems and Optimal Control in Computational Anatomy: 100 Years Since D'Arcy Thompson". Annual Review of Biomedical Engineering. 17 (1): 447–509. doi:10.1146/annurev-bioeng-071114-040601. PMID 26643025.

247. MILLER, MICHAEL I.; TROUVÉ, ALAIN; YOUNES, LAURENT (2006-01-31). "Geodesic Shooting for Computational Anatomy". Journal of mathematical imaging and vision. 24 (2): 209–228. doi:10.1007/s10851-005-3624-0. PMC 2897162 . PMID 20613972.

248. Camion, Vincent; Younes, Laurent (2001-01-01). "Geodesic Interpolating Splines". Proceedings of the Third International Workshop on Energy Minimization Methods in Computer Vision and Pattern Recognition. EMMCVPR '01. London, UK, UK: Springer-Verlag: 513–527. ISBN 3-540-42523-3.

249. Vaillant, M.; Miller, M. I.; Younes, L.; Trouvé, A. (2004-01-01). "Statistics on diffeomorphisms via tangent space representations". NeuroImage. 23 Suppl 1: S161–169. doi:10.1016/j.neuroimage.2004.07.023. PMID 15501085.

250. Marsland, Stephen; McLachlan, Robert (2007-01-01). "A hamiltonian particle method for diffeomorphic image registration". Information Processing in Medical Imaging: Proceedings of the ... Conference. 20: 396–407. PMID 17633716.

251. "L.: Diffeomorphic matching of distributions: A new approach for unlabelled point-sets and sub-manifolds matching". ResearchGate. doi:10.1109/CVPR.2004.1315234. Retrieved 2015-11-25.

252. Beg, M. Faisal; Miller, Michael I.; Trouvé, Alain; Younes, Laurent (2005-02-

01). "Computing Large Deformation Metric Mappings via Geodesic Flows of Diffeomorphisms". International Journal of Computer Vision. 61 (2): 139–157. doi:10.1023/B:VISI.0000043755.93987.aa.

253. Vialard, François-Xavier; Risser, Laurent; Rueckert, Daniel; Cotter, Colin J. (2012-04-01). "Diffeomorphic 3D Image Registration via Geodesic Shooting Using an Efficient Adjoint Calculation". Int. J. Comput. Vision. 97 (2): 229–241. doi:10.1007/s11263-011-0481-8.

254. Glaunès, Joan; Qiu, Anqi; Miller, Michael I.; Younes, Laurent (2008-12-01). "Large Deformation Diffeomorphic Metric Curve Mapping". International Journal of Computer Vision. 80 (3): 317–336. doi:10.1007/s11263-008-0141-9. PMC 2858418 . PMID 20419045.

255. Vaillant, Marc; Glaunès, Joan (2005-01-01). "Surface matching via currents". Proceedings of Information Processing in Medical Imaging (IPMI 2005), number 3565 in Lecture Notes in Computer Science: 381–392.

256. Cao, Yan; Miller, M.I.; Winslow, R.L.; Younes, L. (2005-10-01). "Large deformation diffeomorphic metric mapping of fiber orientations" (PDF). Tenth IEEE International Conference on Computer Vision, 2005. ICCV 2005. 2: 1379–1386 Vol. 2. doi:10.1109/ICCV.2005.132.

257. Cao, Yan; Miller, M.I.; Winslow, R.L.; Younes, L. (2005-09-01). "Large deformation diffeomorphic metric mapping of vector fields". IEEE Transactions on Medical Imaging. 24 (9): 1216–1230. doi:10.1109/TMI.2005.853923. PMID 16156359.

258. Charon, N.; Trouvé, A. (2013-01-01). "The Varifold Representation of Nonoriented Shapes for Diffeomorphic Registration". SIAM Journal on Imaging Sciences. 6 (4): 2547–2580. doi:10.1137/130918885.

259. Bajcsy, R.; Lieberson, R.; Reivich, M. (1983-08-01). "A computerized system for the elastic matching of deformed radiographic images to idealized atlas images". Journal of Computer Assisted Tomography. 7 (4): 618–625. doi:10.1097/00004728-198308000-00008. PMID 6602820.

260. Amit, Yali; Grenander, Ulf; Piccioni, Mauro (1991-06-01). "Structural Image Restoration through Deformable Templates". Journal of the American Statistical Association. 86 (414): 376–387. doi:10.1080/01621459.1991.10475053.

261. Cao, Yan; Miller, M.I.; Mori, Susumu; Winslow, R.L.; Younes, L. (2006-06-01). "Diffeomorphic Matching of Diffusion Tensor Images". Conference on Computer Vision and Pattern Recognition Workshop, 2006. CVPRW '06. 2006: 67–67. doi:10.1109/CVPRW.2006.65. PMC 2920614 . PMID 20711423.

262. Cootes, T. F.; Edwards, G. J.; Taylor, C. J. (1998-06-02). Burkhardt, Hans; Neumann, Bernd, eds. Active appearance models. Lecture Notes in Computer Science. Springer Berlin Heidelberg. pp. 484–498. ISBN 9783540646136.

263. Lian, Nai-Xiang; Davatzikos, Christos (2011-12-01). "Morphological appearance manifolds for group-wise morphometric analysis". Medical Image Analysis. 15 (6): 814–829. doi:10.1016/j.media.2011.06.003. PMC 4392008 . PMID 21873104.

264. Miller, M. I.; Younes, L. (2001-01-01). "Group Actions, Homeomorphisms, and

Matching: A General Framework". International Journal of Computer Vision. 41 (1-2): 61–84. doi:10.1023/A:1011161132514.

265. Trouvé, Alain; Younes, Laurent (2005-01-01). Metamorphoses Through Lie Group Action.

266. Holm, Darryl D.; Trouve, Alain; Younes, Laurent (2008-06-04). "The Euler-Poincare theory of Metamorphosis". arXiv:0806.0870 .

267. Richardson, Casey L.; Younes, Laurent (2014-09-23). "Metamorphosis of Images in Reproducing Kernel Hilbert Spaces". arXiv:1409.6573 .

268. Bookstein, F. L. (1989-01-01). "Principal Warps: Thin-Plate Splines and the Decomposition of Deformations" (PDF). IEEE Transactions on Pattern Analysis and Machine Intelligence. 11 (6): 567–585. doi:10.1109/34.24792.

269. Camion, Vincent; Younes, Laurent (2001-09-03). Figueiredo, Mário; Zerubia, Josiane; Jain, Anil K., eds. Geodesic Interpolating Splines. Lecture Notes in Computer Science. Springer Berlin Heidelberg. pp. 513–527. doi:10.1007/3-540-44745-8_34. ISBN 9783540425236.

270. Glaunes, J.; Trouve, A.; Younes, L. (2004-06-01). "Diffeomorphic matching of distributions: a new approach for unlabelled point-sets and sub-manifolds matching". Proceedings of the 2004 IEEE Computer Society Conference on Computer Vision and Pattern Recognition, 2004. CVPR 2004. 2: II–712–II–718 Vol.2. doi:10.1109/CVPR.2004.1315234.

271. Niethammer, Marc; Huang, Yang; Vialard, François-Xavier (2011-01-01). "Geodesic regression for image time-series". Medical image computing and computer-assisted intervention: MICComputational anatomyI ... International Conference on Medical Image Computing and Computer-Assisted Intervention. 14 (Pt 2): 655–662. PMC 4339064 . PMID 21995085.

272. Trouvé, Alain; Vialard, François-Xavier. "Shape Splines and Stochastic Shape Evolutions: A Second Order Point of View". arXiv:1003.3895 .

273. Fletcher, P.T.; Lu, C.; Pizer, S.M.; Joshi, S. (2004-08-01). "Principal geodesic analysis for the study of nonlinear statistics of shape" (PDF). IEEE Transactions on Medical Imaging. 23 (8): 995–1005. doi:10.1109/TMI.2004.831793.

274. Trouvé, Alain; Vialard, François-Xavier (2012-01-01). "Shape splines and stochastic shape evolutions: A second order point of view". Quarterly of Applied Mathematics. 70 (2): 219–251. doi:10.1090/S0033-569X-2012-01250-4.

275. Miller, Michael; Banerjee, Ayananshu; Christensen, Gary; Joshi, Sarang; Khaneja, Navin; Grenander, Ulf; Matejic, Larissa (1997-06-01). "Statistical methods in computational anatomy". Statistical Methods in Medical Research. 6 (3): 267–299. doi:10.1177/096228029700600305. PMID 9339500.

276. U. Grenander and M. I. Miller (2007-02-08). Pattern Theory: From Representation to Inference. Oxford University Press. ISBN 9780199297061.

277. M. I. Miller and S. Mori and X. Tang and D. Tward and Y. Zhang (2015-02-14). Bayesian Multiple Atlas Deformable Templates. Brain Mapping: An Encyclopedic Reference. Academic Press. ISBN 9780123973160.

278. Srivastava, S.; Miller, M. I.; Grenander, U. (1997-01-01). Byrnes, Christopher I.; Datta, Biswa N.; Martin, Clyde F.; Gilliam, David S., eds. Ergodic Algorithms on Special Euclidean Groups for ATR. Systems & Control: Foundations & Applications. Birkhäuser Boston. pp. 327–350. doi:10.1007/978-1-4612-4120-1_18. ISBN 978-1-4612-8662-2.

279. Kendall, David G. (1989-01-01). "A Survey of the Statistical Theory of Shape". Statistical Science. 4 (2): 87–99. doi:10.1214/ss/1177012582. JSTOR 2245331.

280. Mumford, David (2012-01-01). Zannier, Umberto, ed. The geometry and curvature of shape spaces. Colloquia. Scuola Normale Superiore. pp. 43–53. doi:10.1007/978-88-7642-387-1_4. ISBN 9788876423888.

281. Laurent Younes (2010-05-25). Shapes and Diffeomorphisms (1st ed.). Springer. ISBN 9783642120541.

282. Younes, Laurent (2012-06-01). "Spaces and Manifolds of Shapes in Computer Vision: An Overview". Image Vision Comput. 30 (6-7): 389–397. doi:10.1016/j.imavis.2011.09.009.

283. Ma, Jun; Miller, Michael I.; Younes, Laurent (2010-01-01). "A Bayesian Generative Model for Surface Template Estimation". International Journal of Biomedical Imaging. 2010: 1–14. doi:10.1155/2010/974957. PMC 2946602 . PMID 20885934.

284. Joshi, S.; Davis, Brad; Jomier, B. Matthieu; B, Guido Gerig (2004-01-01). "Unbiased diffeomorphic atlas construction for computational anatomy". NeuroImage. 23: 151–160.

285. Ma, Jun; Miller, Michael I.; Trouvé, Alain; Younes, Laurent (2008-08-01). "Bayesian Template Estimation in Computational Anatomy". NeuroImage. 42 (1): 252–261. doi:10.1016/j.neuroimage.2008.03.056. PMC 2602958 . PMID 18514544.

286. "MriCloud". Retrieved 26 October 2016.

CHAPTER 5

Molecular Cloning

Chapter Contents

- **Introduction**
- **Overview**
- **Steps in Molecular Cloning**
- **Applications of Molecular Cloning**
- **Gene Mapping**

DNA replication involves the copying of each strand of the double helix to give a pair of daughter strands. Replication begins at a specific sequence, called the origin. After initiation begins at an origin sequence, all sequences are replicated no matter what their information. This principle leads to the idea of molecular cloning, or recombinant DNA. Cloning enables the production of a single DNA sequence in large quantities.

INTRODUCTION

Molecular cloning is a set of experimental methods in molecular biology that are used to assemble recombinant DNA molecules and to direct their replication within host organisms. The use of the word cloning refers to the fact that the method involves the replication of one molecule to produce a population of cells with identical DNA molecules. Molecular cloning generally uses DNA sequences from two different organisms: the species that is the source of the DNA to be cloned, and the species that will serve as the living host for replication of the recombinant DNA. Molecular cloning methods are central to many contemporary areas of modern biology and medicine.

In a conventional molecular cloning experiment, the DNA to be cloned is obtained from an organism of interest, then treated with enzymes in the test tube to generate smaller DNA fragments. Subsequently, these fragments are then combined with vector DNA to generate recombinant DNA molecules. The recombinant DNA is then introduced into a host organism (typically an easy-to-grow, benign, laboratory strain of E. coli bacteria). This will generate a population of organisms in which recombinant DNA molecules are replicated along with the host DNA. Because they contain foreign DNA fragments, these are transgenic or genetically modified microorganisms (GMO). This process takes advantage of the fact that a single bacterial cell can be induced to take up and replicate a single recombinant DNA molecule. This single cell can then be expanded exponentially to generate a large amount of bacteria, each of which contain copies of the original recombinant molecule. Thus, both the resulting bacterial population, and the recombinant DNA molecule, are commonly referred to as "clones". Strictly speaking, recombinant DNA refers to DNA molecules, while molecular cloning refers to the experimental methods used to assemble them.

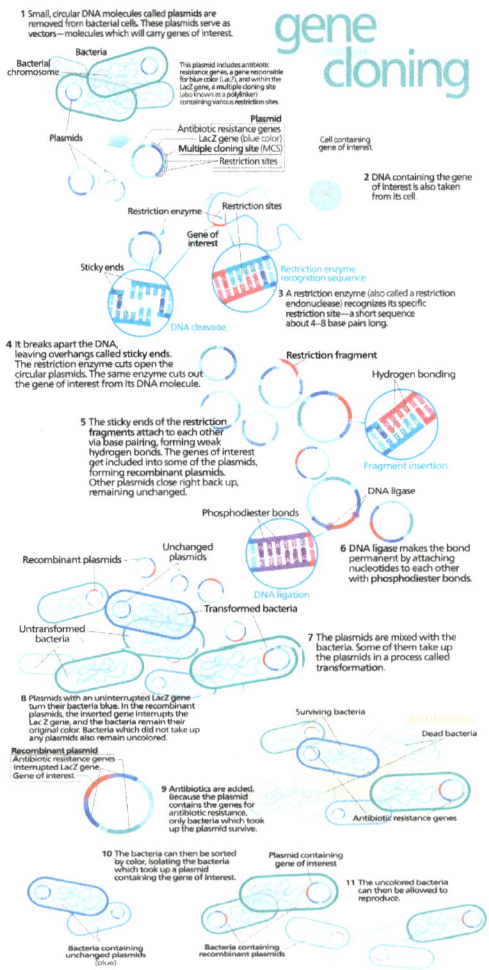

Diagram of molecular cloning using bacteria and plasmids.

History of Molecular Cloning

Prior to the 1970s, our understanding of genetics and molecular biology was severely hampered by an inability to isolate and study individual genes from complex organisms. This changed dramatically with the advent of molecular cloning methods. Microbiologists, seeking to understand the molecular mechanisms through which bacteria restricted the growth of bacteriophage, isolated restriction endonucleases, enzymes that could cleave DNA molecules only when specific DNA sequences were encountered. They showed that restriction enzymes cleaved chromosome-length DNA molecules at specific locations, and that specific sections of the larger molecule could be purified by size fractionation. Using a second enzyme, DNA ligase, fragments generated by restriction enzymes could be joined in new combinations, termed recombinant DNA. By recombining DNA segments of interest with vector DNA, such as bacteriophage or plasmids, which naturally replicate inside bacteria, large quantities of purified recombinant DNA molecules could be produced in bacterial cultures. The first recombinant DNA molecules were generated and studied in 1972.

OVERVIEW

Molecular cloning takes advantage of the fact that the chemical structure of DNA is fundamentally the same in all living organisms. Therefore, if any segment of DNA from any organism is inserted into a DNA segment containing the molecular sequences required for DNA replication, and the resulting recombinant DNA is introduced into the organism from which the replication sequences were obtained, then the foreign DNA will be replicated along with the host cell's DNA in the transgenic organism

Molecular cloning is similar to polymerase chain reaction (PCR) in that it permits the replication of DNA sequence. The fundamental difference between the two methods is that molecular cloning involves replication of the DNA in a living microorganism, while PCR replicates DNA in an *in vitro* solution, free of living cells.

STEPS IN MOLECULAR CLONING

In standard molecular cloning experiments, the cloning of any DNA fragment essentially involves seven steps: (1) Choice of host organism and cloning vector, (2) Preparation of vector DNA, (3) Preparation of DNA to be cloned, (4) Creation of recombinant DNA, (5) Introduction of recombinant DNA into host organism, (6) Selection of organisms containing recombinant DNA, (7) Screening for clones with desired DNA inserts and biological properties.

Although the detailed planning of the cloning can be done in any text editor, together with online utilities for e.g. PCR primer design, dedicated software exist for the purpose. Software for the purpose include for example ApE (open source), DNAStrider (open source), Serial Cloner (gratis) and Collagene (open source).

Choice of Host Organism and Cloning Vector

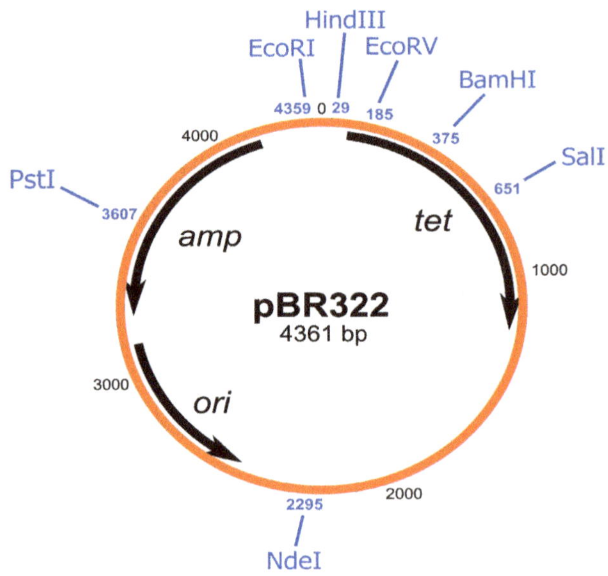

Diagram of a commonly used cloning plasmid; pBR322. It's a circular piece of DNA 4361 bases long. Two antibiotic resistancegenes are present, conferring resistance to ampicillin and tetracycline, and an origin of replication that the host uses to replicate the DNA.

Although a very large number of host organisms and molecular cloning vectors are in use, the great majority of molecular cloning experiments begin with a laboratory strain of the bacterium *E. coli* (*Escherichia coli*) and a plasmid cloning vector. *E. coli* and plasmid vectors are in common use because they are technically sophisticated, versatile, widely available, and offer rapid growth of recombinant organisms with minimal equipment. If the DNA to be cloned is exceptionally large (hundreds of thousands to millions of base pairs), then a bacterial artificial chromosome or yeast artificial chromosome vector is often chosen.

Specialized applications may call for specialized host-vector systems. For example, if the experimentalists wish to harvest a particular protein from the recombinant organism, then an expression vector is chosen that contains appropriate signals for transcription and translation in the desired host organism. Alternatively, if replication of the DNA in different species is desired (for example, transfer of DNA from bacteria to plants), then a multiple host range vector (also termed shuttle vector) may be selected. In practice, however, specialized molecular cloning experiments usually begin with cloning into a bacterial plasmid, followed by subcloning into a specialized vector.

Whatever combination of host and vector are used, the vector almost always contains four DNA segments that are critically important to its function and experimental utility:

- DNA *replication origin* is necessary for the vector (and its linked recombinant sequences) to replicate inside the host organism

- one or more unique *restriction endonuclease recognition sites* to serves as sites where foreign DNA may be introduced
- a *selectable genetic marker* gene that can be used to enable the survival of cells that have taken up vector sequences
- a *tag* gene that can be used to screen for cells containing the foreign DNA

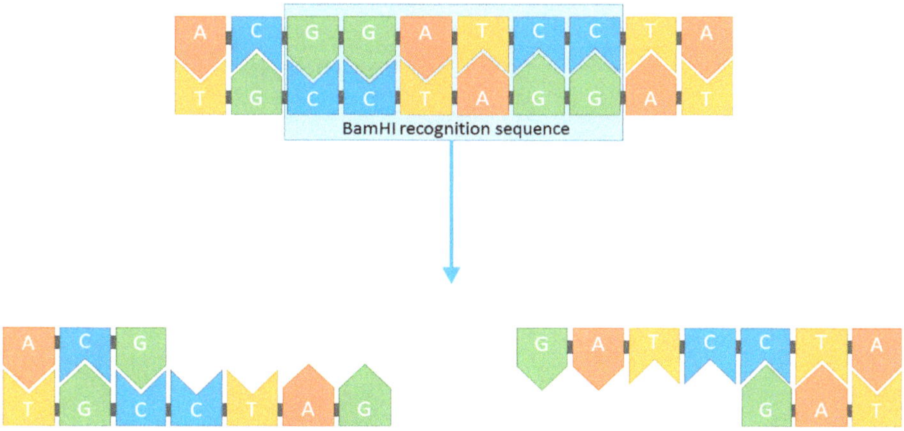

Cleavage of a DNA sequence containing the BamHI restriction site. The DNA is cleaved at the palindromic sequence to produce 'sticky ends'.

Preparation of Vector DNA

The cloning vector is treated with a restriction endonuclease to cleave the DNA at the site where foreign DNA will be inserted. The restriction enzyme is chosen to generate a configuration at the cleavage site that is compatible with the ends of the foreign DNA (see DNA end). Typically, this is done by cleaving the vector DNA and foreign DNA with the same restriction enzyme, for example EcoRI. Most modern vectors contain a variety of convenient cleavage sites that are unique within the vector molecule (so that the vector can only be cleaved at a single site) and are located within a gene (frequently beta-galactosidase) whose inactivation can be used to distinguish recombinant from non-recombinant organisms at a later step in the process. To improve the ratio of recombinant to non-recombinant organisms, the cleaved vector may be treated with an enzyme (alkaline phosphatase) that dephosphorylates the vector ends. Vector molecules with dephosphorylated ends are unable to replicate, and replication can only be restored if foreign DNA is integrated into the cleavage site.

Preparation of DNA to be Cloned

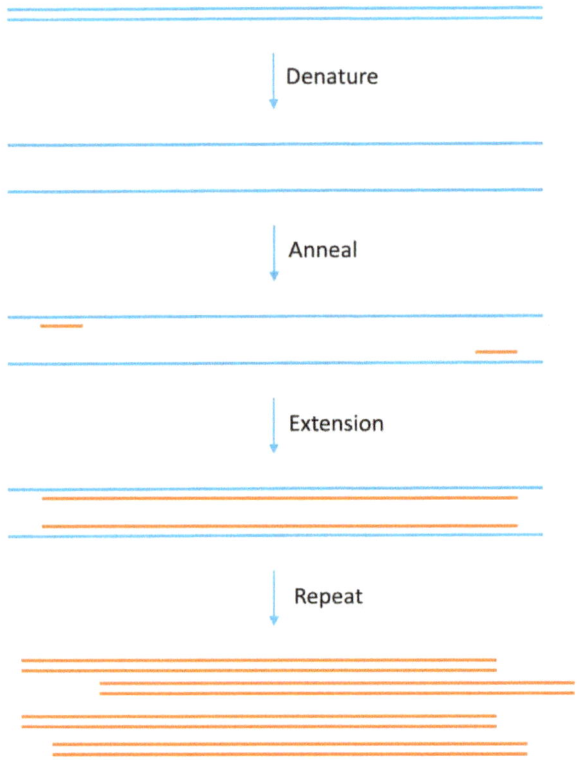

DNA for cloning is most commonly produced using PCR. Template DNA is mixed with bases (the building blocks of DNA), primers (short pieces of complementary single stranded DNA) and a DNA polymerase enzyme that builds the DNA chain. The mix goes through cycles of heating and cooling to produce large quantities of copied DNA.

For cloning of genomic DNA, the DNA to be cloned is extracted from the organism of interest. Virtually any tissue source can be used (even tissues from extinct animals), as long as the DNA is not extensively degraded. The DNA is then purified using simple methods to remove contaminating proteins (extraction with phenol), RNA (ribonuclease) and smaller molecules (precipitation and/or chromatography). Polymerase chain reaction (PCR) methods are often used for amplification of specific DNA or RNA (RT-PCR) sequences prior to molecular cloning.

DNA for cloning experiments may also be obtained from RNA using reverse transcriptase (complementary DNA or cDNA cloning), or in the form of synthetic DNA (artificial gene synthesis). cDNA cloning is usually used to obtain clones representative of the mRNA population of the cells of interest, while synthetic DNA is used to obtain any precise sequence defined by the designer.

The purified DNA is then treated with a restriction enzyme to generate fragments with ends capable of being linked to those of the vector. If necessary, short double-stranded segments of DNA (*linkers*) containing desired restriction sites may be added to create end structures that are compatible with the vector.

Creation of Recombinant DNA with DNA Ligase

The creation of recombinant DNA is in many ways the simplest step of the molecular cloning process. DNA prepared from the vector and foreign source are simply mixed together at appropriate concentrations and exposed to an enzyme (DNA ligase) that covalently links the ends together. This joining reaction is often termed ligation. The resulting DNA mixture containing randomly joined ends is then ready for introduction into the host organism.

DNA ligase only recognizes and acts on the ends of linear DNA molecules, usually resulting in a complex mixture of DNA molecules with randomly joined ends. The desired products (vector DNA covalently linked to foreign DNA) will be present, but other sequences (e.g. foreign DNA linked to itself, vector DNA linked to itself and higher-order combinations of vector and foreign DNA) are also usually present. This complex mixture is sorted out in subsequent steps of the cloning process, after the DNA mixture is introduced into cells.

Introduction of Recombinant DNA into Host Organism

The DNA mixture, previously manipulated in vitro, is moved back into a living cell, referred to as the host organism. The methods used to get DNA into cells are varied, and the name applied to this step in the molecular cloning process will often depend upon the experimental method that is chosen (e.g. transformation, transduction, transfection, electroporation).

When microorganisms are able to take up and replicate DNA from their local environment, the process is termed transformation, and cells that are in a physiological state such that they can take up DNA are said to be competent. In mammalian cell culture, the analogous process of introducing DNA into cells is commonly termed transfection. Both transformation and transfection usually require preparation of the cells through a special growth regime and chemical treatment process that will vary with the specific species and cell types that are used.

Electroporation uses high voltage electrical pulses to translocate DNA across the cell membrane (and cell wall, if present). In contrast, transduction involves the packaging of DNA into virus-derived particles, and using these virus-like particles to introduce the encapsulated DNA into the cell through a process resembling viral infection. Although electroporation and transduction are highly specialized methods, they may be the most efficient methods to move DNA into cells.

Selection of Organisms Containing Vector Sequences

Whichever method is used, the introduction of recombinant DNA into the chosen host organism is usually a low efficiency process; that is, only a small fraction of the cells will actually take up DNA. Experimental scientists deal with this issue through a step of artificial genetic selection, in which cells that have not taken up DNA are selectively killed, and only those cells that can actively replicate DNA containing the selectable marker gene encoded by the vector are able to survive.

When bacterial cells are used as host organisms, the selectable marker is usually a gene that confers resistance to an antibiotic that would otherwise kill the cells, typically ampicillin. Cells harboring the plasmid will survive when exposed to the antibiotic, while those that have failed to take up plasmid sequences will die. When mammalian cells (e.g. human or mouse cells) are used, a similar strategy is used, except that the marker gene (in this case typically encoded as part of the kanMX cassette) confers resistance to the antibiotic Geneticin.

Screening for Clones with Desired DNA Inserts and Biological Properties

Modern bacterial cloning vectors (e.g. pUC19 and later derivatives including the pGEM vectors) use the blue-white screening system to distinguish colonies (clones) of transgenic cells from those that contain the parental vector (i.e. vector DNA with no recombinant sequence inserted). In these vectors, foreign DNA is inserted into a sequence that encodes an essential part of beta-galactosidase, an enzyme whose activity results in formation of a blue-colored colony on the culture medium that is used for this work. Insertion of the foreign DNA into the beta-galactosidase coding sequence disables the function of the enzyme, so that colonies containing transformed DNA remain colorless (white). Therefore, experimentalists are easily able to identify and conduct further studies on transgenic bacterial clones, while ignoring those that do not contain recombinant DNA.

The total population of individual clones obtained in a molecular cloning experiment is often termed a DNA library. Libraries may be highly complex (as when cloning complete genomic DNA from an organism) or relatively simple (as when moving a previously cloned DNA fragment into a different plasmid), but it is almost always necessary to examine a number of different clones to be sure that the desired DNA construct is obtained. This may be accomplished through a very wide range of experimental methods, including the use of nucleic acid hybridizations, antibody probes, polymerase chain reaction, restriction fragment analysis and/or DNA sequencing.

APPLICATIONS OF MOLECULAR CLONING

Molecular cloning provides scientists with an essentially unlimited quantity of any individual DNA segments derived from any genome. This material can be used for a wide range of purposes, including those in both basic and applied biological science. A few of the more important applications are summarized here.

Genome Organization and Gene Expression

Molecular cloning has led directly to the elucidation of the complete DNA sequence of the genomes of a very large number of species and to an exploration of genetic diversity within individual species, work that has been done mostly by determining the DNA sequence of large numbers of randomly cloned fragments of the genome, and assembling the overlapping sequences.

At the level of individual genes, molecular clones are used to generate probes that are used for examining how genes are expressed, and how that expression is related to other processes in biology, including the metabolic environment, extracellular signals, development, learning, senescence and cell death. Cloned genes can also provide tools to examine the biological function and importance of individual genes, by allowing investigators to inactivate the genes, or make more subtle mutations using regional mutagenesis or site-directed mutagenesis.

Production of Recombinant Proteins

Obtaining the molecular clone of a gene can lead to the development of organisms that produce the protein product of the cloned genes, termed a recombinant protein. In practice, it is frequently more difficult to develop an organism that produces an active form of the recombinant protein in desirable quantities than it is to clone the gene. This is because the molecular signals for gene expression are complex and variable, and because protein folding, stability and transport can be very challenging.

Many useful proteins are currently available as recombinant products. These include--(1) medically useful proteins whose administration can correct a defective or poorly expressed gene (e.g. recombinant factor VIII, a blood-clotting factor deficient in some forms of hemophilia, and recombinant insulin, used to treat some forms of diabetes), (2) proteins that can be administered to assist in a life-threatening emergency (e.g. tissue plasminogen activator, used to treat strokes), (3) recombinant subunit vaccines, in which a purified protein can be used to immunize patients against infectious diseases, without exposing them to the infectious agent itself (e.g. hepatitis B vaccine), and (4) recombinant proteins as standard matcrial for diagnostic laboratory tests.

Transgenic Organisms

Once characterized and manipulated to provide signals for appropriate expression, cloned genes may be inserted into organisms, generating transgenic organisms, also termed genetically modified organisms (GMOs). Although most GMOs are generated for purposes of basic biological research (see for example, transgenic mouse), a number of GMOs have been developed for commercial use, ranging from animals and plants that produce pharmaceuticals or other compounds (pharming), herbicide-resistant crop plants, and fluorescent tropical fish (GloFish) for home entertainment.

Gene Therapy

Gene therapy involves supplying a functional gene to cells lacking that function, with the aim of correcting a genetic disorder or acquired disease. Gene therapy can be broadly divided into two categories. The first is alteration of germ cells, that is, sperm or eggs, which results in a permanent genetic change for the whole organism and subsequent generations. This "germ line gene therapy" is considered by many to be unethical in human beings. The second type of gene therapy, "somatic cell gene therapy", is analogous to an organ transplant. In this case, one or more specific tissues are targeted by direct treatment or by removal of the tissue, addition of the therapeutic gene or genes in the laboratory, and return of the treated cells to the patient. Clinical trials of somatic cell gene therapy began in the late 1990s, mostly for the treatment of cancers and blood, liver, and lung disorders.

Despite a great deal of publicity and promises, the history of human gene therapy has been characterized by relatively limited success. The effect of introducing a gene into cells often promotes only partial and/or transient relief from the symptoms of the disease being treated. Some gene therapy trial patients have suffered adverse consequences of the treatment itself, including deaths. In some cases, the adverse effects result from disruption of essential genes within the patient's genome by insertional inactivation. In others, viral vectors used for gene therapy have been contaminated with infectious virus. Nevertheless, gene therapy is still held to be a promising future area of medicine, and is an area where there is a significant level of research and development activity.

Cloning

In biology, cloning is the process of producing similar populations of genetically identical individuals that occurs in nature when organisms such as bacteria, insects or plants reproduce asexually. Cloning in biotechnology refers to processes used to create copies of DNA fragments (molecular cloning), cells (cell cloning), or organisms. The term also refers to the production of multiple copies of a product such as digital media or software.

The term clone, invented by J. B. S. Haldane, is derived from the Ancient Greek word κλών klōn, "twig", referring to the process whereby a new plant can be created from a twig. In horticulture, the spelling clon was used until the twentieth century; the final e came into use to indicate the vowel is a "long o" instead of a "short o". Since the term entered the popular lexicon in a more general context, the spelling clone has been used exclusively.

In botany, the term lusus was traditionally used.:21, 43

Many organisms, including aspen trees, reproduce by cloning.

Natural Cloning

Cloning is a natural form of reproduction that has allowed life forms to spread for more than 50 thousand years. It is the reproduction method used by plants, fungi, and bacteria, and is also the way that clonal colonies reproduce themselves. Examples of these organisms include blueberry plants, hazel trees, the Pando trees, the Kentucky coffeetree, *Myrica*s, and the American sweetgum.

Molecular Cloning

Molecular cloning refers to the process of making multiple molecules. Cloning is commonly used to amplify DNA fragments containing whole genes, but it can also be used to amplify any DNA sequence such as promoters, non-coding sequences and randomly fragmented DNA. It is used in a wide array of biological experiments and practical applications ranging from genetic fingerprinting to large scale protein production. Occasionally, the term cloning is misleadingly used to refer to the identification of the chromosomal location of a gene associated with a particular phenotype of interest, such as in positional cloning. In practice, localization of the gene to a chromosome or genomic region does not necessarily enable one to isolate or amplify the relevant genomic sequence. To amplify any DNA sequence in a living organism, that sequence must be linked to an origin of replication, which is a sequence of DNA capable of directing the propagation of itself and any linked sequence. However, a number of other features are needed, and a variety of specialised cloning vectors (small piece of DNA into which a foreign DNA fragment can be inserted) exist that allow protein production, affinity tagging, single stranded RNA or DNA production and a host of other molecular biology tools.

Cloning of any DNA fragment essentially involves four steps

- fragmentation - breaking apart a strand of DNA
- ligation - gluing together pieces of DNA in a desired sequence
- transfection - inserting the newly formed pieces of DNA into cells
- screening/selection - selecting out the cells that were successfully transfected with the new DNA

Although these steps are invariable among cloning procedures a number of alternative routes can be selected; these are summarized as a *cloning strategy*.

Initially, the DNA of interest needs to be isolated to provide a DNA segment of suitable size. Subsequently, a ligation procedure is used where the amplified fragment is inserted into a vector (piece of DNA). The vector (which is frequently circular) is linearised using restriction enzymes, and incubated with the fragment of interest under appropriate conditions with an enzyme called DNA ligase. Following ligation the vector with the insert of interest is transfected into cells. A number of alternative techniques are available, such as chemical sensitivation of cells, electroporation, optical injection and biolistics. Finally, the transfected cells are cultured. As the aforementioned procedures are of particularly low efficiency, there is a need to identify the cells that have been successfully transfected with the vector construct containing the desired insertion sequence in the required orientation. Modern cloning vectors include selectable antibiotic resistance markers, which allow only cells in which the vector has been transfected, to grow. Additionally, the cloning vectors may contain colour selection markers, which provide blue/white screening (alpha-factor complementation) on X-gal medium. Nevertheless, these selection steps do not absolutely guarantee that the DNA insert is present in the cells obtained. Further investigation of the resulting colonies must be required to confirm that cloning was successful. This may be accomplished by means of PCR, restriction fragment analysis and/or DNA sequencing.

Cell Cloning

Cloning Unicellular Organisms

Cloning a cell means to derive a population of cells from a single cell. In the case of unicellular organisms such as bacteria and yeast, this process is remarkably simple and essentially only requires the inoculation of the appropriate medium. However, in the case of cell cultures from multi-cellular organisms, cell cloning is an arduous task as these cells will not readily grow in standard media.

A useful tissue culture technique used to clone distinct lineages of cell lines involves the use of cloning rings (cylinders). In this technique a single-cell suspension of cells that have been exposed to a mutagenic agent or drug used to drive selection is plated at high dilution to create isolated colonies, each arising from a single and potentially clonal distinct cell. At an early growth stage when colonies consist of only a few cells, sterile polystyrene rings (cloning rings), which have been dipped in grease, are placed over an individual colony and a small amount of trypsin is added. Cloned cells are collected from inside the ring and transferred to a new vessel for further growth.

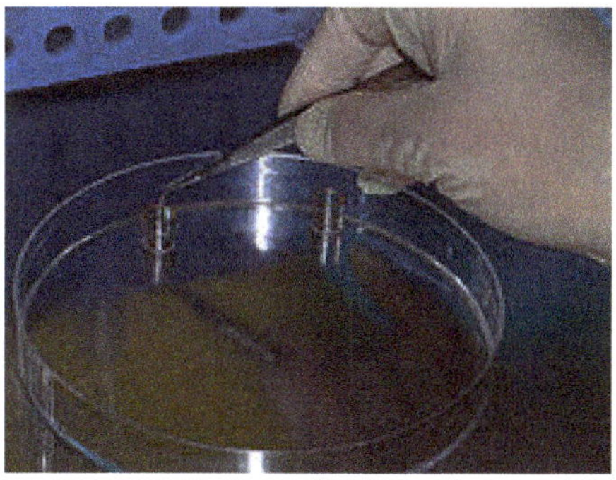

Cloning cell-line colonies using cloning rings

Cloning Stem Cells

Somatic-cell nuclear transfer, known as SCNT, can also be used to create embryos for research or therapeutic purposes. The most likely purpose for this is to produce embryos for use in stem cell research. This process is also called "research cloning" or "therapeutic cloning." The goal is not to create cloned human beings (called "reproductive cloning"), but rather to harvest stem cells that can be used to study human development and to potentially treat disease. While a clonal human blastocyst has been created, stem cell lines are yet to be isolated from a clonal source.

Therapeutic cloning is achieved by creating embryonic stem cells in the hopes of treating diseases such as diabetes and Alzheimer's. The process begins by removing the nucleus (containing the DNA) from an egg cell and inserting a nucleus from the adult cell to be cloned. In the case of someone with Alzheimer's disease, the nucleus from a skin cell of that patient is placed into an empty egg. The reprogrammed cell begins to develop into an embryo because the egg reacts with the transferred nucleus. The embryo will become genetically identical to the patient. The embryo will then form a blastocyst which has the potential to form/become any cell in the body.

The reason why SCNT is used for cloning is because somatic cells can be easily acquired and cultured in the lab. This process can either add or delete specific genomes of farm animals. A key point to remember is that cloning is achieved when the oocyte maintains its normal functions and instead of using sperm and egg genomes to replicate, the oocyte is inserted into the donor's somatic cell nucleus. The oocyte will react on the somatic cell nucleus, the same way it would on sperm cells.

The process of cloning a particular farm animal using SCNT is relatively the same for all animals. The first step is to collect the somatic cells from the animal that will be cloned. The somatic cells could be used immediately or stored in the laboratory for later use. The hardest part of SCNT is removing maternal DNA from an oocyte at metaphase II. Once this has been done, the somatic nucleus can be inserted into an egg cytoplasm. This creates a one-cell embryo. The grouped somatic cell and egg cytoplasm are then introduced to an electrical current. This energy will hopefully allow the cloned embryo to begin development. The successfully developed embryos are then placed in surrogate recipients, such as a cow or sheep in the case of farm animals.

SCNT is seen as a good method for producing agriculture animals for food consumption. It successfully cloned sheep, cattle, goats, and pigs. Another benefit is SCNT is seen as a solution to clone endangered species that are on the verge of going extinct. However, stresses placed on both the egg cell and the introduced nucleus can be enormous, which led to a high loss in resulting cells in early research. For example, the cloned sheep Dolly was born after 277 eggs were used for SCNT, which created 29 viable embryos. Only three of these embryos survived until birth, and only one survived to adulthood. As the procedure could not be automated, and had to be performed manually under a microscope, SCNT was very resource intensive. The biochemistry involved in reprogramming the differentiated somatic cell nucleus and activating the recipient egg was also far from being well-understood. However, by 2014 researchers were reporting cloning success rates of seven to eight out of ten and in 2016, a Korean Company Sooam Biotech was reported to be producing 500 cloned embryos per day.

In SCNT, not all of the donor cell›s genetic information is transferred, as the donor cell›s mitochondria that contain their own mitochondrial DNA are left behind. The resulting hybrid cells retain those mitochondrial structures which originally belonged to the egg. As a consequence, clones such as Dolly that are born from SCNT are not perfect copies of the donor of the nucleus.

Organism Cloning

Organism cloning (also called reproductive cloning) refers to the procedure of creating a new multicellular organism, genetically identical to another. In essence this form of cloning is an asexual method of reproduction, where fertilization or inter-gamete contact does not take place. Asexual reproduction is a naturally occurring phenomenon in many species, including most plants (see vegetative reproduction) and some insects. Scientists have made some major achievements with cloning, including the asexual reproduction of sheep and cows. There is a lot of ethical debate over whether or not cloning should be used. However, cloning, or asexual propagation, has been common practice in the horticultural world for hundreds of years.

Horticultural

The term *clone* is used in horticulture to refer to descendants of a single plant which were produced by vegetative reproduction or apomixis. Many horticultural plant cultivars are clones, having been derived from a single individual, multiplied by some process other than sexual reproduction. As an example, some European cultivars of grapes represent clones that have been propagated for over two millennia. Other examples are potato and banana. Grafting can be regarded as cloning, since all the shoots and branches coming from the graft are genetically a clone of a single individual, but this particular kind of cloning has not come under ethical scrutiny and is generally treated as an entirely different kind of operation.

Many trees, shrubs, vines, ferns and other herbaceous perennials form clonal colonies naturally. Parts of an individual plant may become detached by fragmentation and grow on to become separate clonal individuals. A common example is in the vegetative reproduction of moss and liverwort gametophyte clones by means of gemmae. Some vascular plants e.g. dandelion and certain viviparous grasses also form seeds asexually, termed apomixis, resulting in clonal populations of genetically identical individuals.

Parthenogenesis

Clonal derivation exists in nature in some animal species and is referred to as parthenogenesis (reproduction of an organism by itself without a mate). This is an asexual form of reproduction that is only found in females of some insects, crustaceans, nematodes, fish (for example the hammerhead shark), the Komodo dragon and lizards. The growth and development occurs without fertilization by a male. In plants, parthenogenesis means the development of an embryo from an unfertilized egg cell, and is a component process of apomixis. In species that use the XY sex-determination system, the offspring will always be female. An example is the little fire ant (*Wasmannia auropunctata*), which is native to Central and South America but has spread throughout many tropical environments.

Artificial Cloning of Organisms

Artificial cloning of organisms may also be called *reproductive cloning.*

First Moves

Hans Spemann, a German embryologist was awarded a Nobel Prize in Physiology or Medicine in 1935 for his discovery of the effect now known as embryonic induction, exercised by various parts of the embryo, that directs the development of groups of cells into particular tissues and organs. In 1928 he and his student, Hilde Mangold, were the first to perform somatic-cell nuclear transfer using amphibian embryos – one of the first moves towards cloning.

Methods

Reproductive cloning generally uses "somatic cell nuclear transfer" (SCNT) to create animals that are genetically identical. This process entails the transfer of a nucleus from a donor adult cell (somatic cell) to an egg from which the nucleus has been removed, or to a cell from a blastocyst from which the nucleus has been removed. If the egg begins to divide normally it is transferred into the uterus of the surrogate mother. Such clones are not strictly identical since the somatic cells may contain mutations in their nuclear DNA. Additionally, the mitochondria in the cytoplasm also contains DNA and during SCNT this mitochondrial DNA is wholly from the cytoplasmic donor's egg, thus the mitochondrial genome is not the same as that of the nucleus donor cell from which it was produced. This may have important implications for cross-species nuclear transfer in which nuclear-mitochondrial incompatibilities may lead to death.

Artificial *embryo splitting* or *embryo twinning*, a technique that creates monozygotic twins from a single embryo, is not considered in the same fashion as other methods of cloning. During that procedure, an donor embryo is split in two distinct embryos, that can then be transferred via embryo transfer. It is optimally performed at the 6- to 8-cell stage, where it can be used as an expansion of IVF to increase the number of available embryos. If both embryos are successful, it gives rise to monozygotic (identical) twins.

Dolly the Sheep

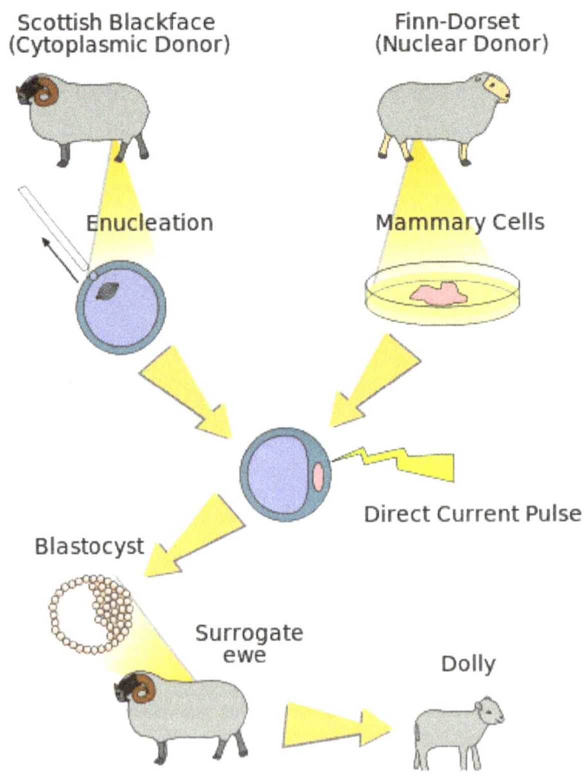

Dolly clone

Dolly, a Finn-Dorset ewe, was the first mammal to have been successfully cloned from an adult somatic cell. Dolly was formed by taking a cell from the udder of her 6-year old biological mother. Dolly›s embryo was created by taking the cell and inserting it into a sheep ovum. It took 434 attempts before an embryo was successful. The embryo was then placed inside a female sheep that went through a normal pregnancy. She was cloned at the Roslin Institute in Scotland by British scientists Sir Ian Wilmut and Keith Campbell and lived there from her birth in 1996 until her death in 2003 when she was six. She was born on 5 July 1996 but not announced to the world until 22 February 1997. Her stuffed remains were placed at Edinburgh's Royal Museum, part of the National Museums of Scotland.

Dolly was publicly significant because the effort showed that genetic material from a specific adult cell, programmed to express only a distinct subset of its genes, can be reprogrammed to grow an entirely new organism. Before this demonstration, it had been shown by John Gurdon that nuclei from differentiated cells could give rise to an entire organism after transplantation into an enucleated egg. However, this concept was not yet demonstrated in a mammalian system.

The first mammalian cloning (resulting in Dolly the sheep) had a success rate of 29 embryos per 277 fertilized eggs, which produced three lambs at birth, one of which lived. In a bovine experiment involving 70 cloned calves, one-third of the calves died young. The first successfully cloned horse, Prometea, took 814 attempts. Notably, although the first clones were frogs, no adult cloned frog has yet been produced from a somatic adult nucleus donor cell.

There were early claims that Dolly the sheep had pathologies resembling accelerated aging. Scientists speculated that Dolly's death in 2003 was related to the shortening of telomeres, DNA-protein complexes that protect the end of linear chromosomes. However, other researchers, including Ian Wilmut who led the team that successfully cloned Dolly, argue that Dolly's early death due to respiratory infection was unrelated to deficiencies with the cloning process. This idea that the nuclei have not irreversibly aged was shown in 2013 to be true for mice.

Dolly was named after performer Dolly Parton because the cells cloned to make her were from a mammary gland cell, and Parton is known for her ample cleavage.

Species Cloned

The modern cloning techniques involving nuclear transfer have been successfully performed on several species. Notable experiments include:

- Tadpole: (1952) Robert Briggs and Thomas J. King had successfully cloned northern leopard frogs: thirty-five complete embryos and twenty-seven tadpoles from one-hundred and four successful nuclear transfers.
- Carp: (1963) In China, embryologist Tong Dizhou produced the world's first cloned fish by inserting the DNA from a cell of a male carp into an egg from a female carp. He published the findings in a Chinese science journal.
- Mice: (1986) A mouse was successfully cloned from an early embryonic cell. Soviet scientists Chaylakhyan, Veprencev, Sviridova, and Nikitin had the mouse "Masha" cloned. Research was published in the magazine "Biofizika" volume XXXII, issue 5 of 1987.
- Sheep: Marked the first mammal being cloned (1984) from early embryonic cells by Steen Willadsen. Megan and Morag cloned from differentiated embryonic cells in June 1995 and Dolly the sheep from a somatic cell in 1996.
- Rhesus monkey: Tetra (January 2000) from embryo splitting

- Pig: the first cloned pigs (March 2000). By 2014, BGI in China was producing 500 cloned pigs a year to test new medicines.
- Gaur: (2001) was the first endangered species cloned.
- Cattle: Alpha and Beta (males, 2001) and (2005) Brazil
- Cat: CopyCat "CC" (female, late 2001), Little Nicky, 2004, was the first cat cloned for commercial reasons
- Rat: Ralph, the first cloned rat (2003)
- Mule: Idaho Gem, a john mule born 4 May 2003, was the first horse-family clone.
- Horse: Prometea, a Haflinger female born 28 May 2003, was the first horse clone.
- Dog: Snuppy, a male Afghan hound was the first cloned dog (2005).
- Wolf: Snuwolf and Snuwolffy, the first two cloned female wolves (2005).
- Water buffalo: Samrupa was the first cloned water buffalo. It was born on 6 February 2009, at India›s Karnal National Diary Research Institute but died five days later due to lung infection.
- Pyrenean ibex (2009) was the first extinct animal to be cloned back to life; the clone lived for seven minutes before dying of lung defects.
- Camel: (2009) Injaz, is the first cloned camel.
- Pashmina goat: (2012) Noori, is the first cloned pashmina goat. Scientists at the faculty of veterinary sciences and animal husbandry of Sher-e-Kashmir University of Agricultural Sciences and Technology of Kashmir successfully cloned the first Pashmina goat (Noori) using the advanced reproductive techniques under the leadership of Riaz Ahmad Shah.
- Gastric brooding frog: (2013) The gastric brooding frog, *Rheobatrachus silus*, thought to have been extinct since 1983 was cloned in Australia, although the embryos died after a few days.

Human Cloning

Human cloning is the creation of a genetically identical copy of a human. The term is generally used to refer to artificial human cloning, which is the reproduction of human cells and tissues. It does not refer to the natural conception and delivery of identical twins. The possibility of human cloning has raised controversies. These ethical concerns have prompted several nations to pass legislature regarding human cloning and its legality.

Two commonly discussed types of theoretical human cloning are *therapeutic cloning* and *reproductive cloning*. Therapeutic cloning would involve cloning cells from a human for use in medicine and transplants, and is an active area of research, but is not in medical practice anywhere in the world, as of 2014. Two common methods of therapeutic cloning that are being researched are somatic-cell nuclear transfer and, more recently, pluripotent stem cell induction. Reproductive cloning would involve making an entire cloned human, instead of just specific cells or tissues.

Ethical Issues of Cloning

There are a variety of ethical positions regarding the possibilities of cloning, especially human cloning. While many of these views are religious in origin, the questions raised by cloning are faced by secular perspectives as well. Perspectives on human cloning are theoretical, as human therapeutic and reproductive cloning are not commercially used; animals are currently cloned in laboratories and in livestock production.

Advocates support development of therapeutic cloning in order to generate tissues and whole organs to treat patients who otherwise cannot obtain transplants,to avoid the need for immunosuppressive drugs, and to stave off the effects of aging. Advocates for reproductive cloning believe that parents who cannot otherwise procreate should have access to the technology.

Opponents of cloning have concerns that technology is not yet developed enough to be safe and that it could be prone to abuse (leading to the generation of humans from whom organs and tissues would be harvested), as well as concerns about how cloned individuals could integrate with families and with society at large.

Religious groups are divided, with some opposing the technology as usurping "God›s place" and, to the extent embryos are used, destroying a human life; others support therapeutic cloning›s potential life-saving benefits.

Cloning of animals is opposed by animal-groups due to the number of cloned animals that suffer from malformations before they die, and while food from cloned animals has been approved by the US FDA, its use is opposed by groups concerned about food safety.

Cloning Extinct and Endangered Species

Cloning, or more precisely, the reconstruction of functional DNA from extinct species has, for decades, been a dream. Possible implications of this were dramatized in the 1984 novel *Carnosaur* and the 1990 novel *Jurassic Park*. The best current cloning techniques have an average success rate of 9.4 percent (and as high as 25 percent) when working with familiar spccics such as mice, while cloning wild animals is usually less than 1 percent successful. Several tissue banks have come into existence, including the "Frozen Zoo" at the San Diego Zoo, to store frozen tissue from the world's rarest and most endangered species.

In 2001, a cow named Bessie gave birth to a cloned Asian gaur, an endangered species, but the calf died after two days. In 2003, a banteng was successfully cloned, followed by three African wildcats from a thawed frozen embryo. These successes provided hope that similar techniques (using surrogate mothers of another species) might be used to clone extinct species. Anticipating this possibility, tissue samples from the last *bucardo* (Pyrenean ibex) were frozen in liquid nitrogen immediately after it died in 2000. Researchers are also considering cloning endangered species such as the giant panda and cheetah.

In 2002, geneticists at the Australian Museum announced that they had replicated DNA of the thylacine (Tasmanian tiger), at the time extinct for about 65 years, using polymerase chain reaction. However, on 15 February 2005 the museum announced that it was stopping the project after tests showed the specimens› DNA had been too badly degraded by the (ethanol) preservative. On 15 May 2005 it was announced that the thylacine project would be revived, with new participation from researchers in New South Wales and Victoria.

In January 2009, for the first time, an extinct animal, the Pyrenean ibex mentioned above was cloned, at the Centre of Food Technology and Research of Aragon, using the preserved frozen cell nucleus of the skin samples from 2001 and domestic goat egg-cells. The ibex died shortly after birth due to physical defects in its lungs.

One of the most anticipated targets for cloning was once the woolly mammoth, but attempts to extract DNA from frozen mammoths have been unsuccessful, though a joint Russo-Japanese team is currently working toward this goal. In January 2011, it was reported by Yomiuri Shimbun that a team of scientists headed by Akira Iritani of Kyoto University had built upon research by Dr. Wakayama, saying that they will extract DNA from a mammoth carcass that had been preserved in a Russian laboratory and insert it into the egg cells of an African elephant in hopes of producing a mammoth embryo. The researchers said they hoped to produce a baby mammoth within six years. It was noted, however that the result, if possible, would be an elephant-mammoth hybrid rather than a true mammoth.Another problem is the survival of the reconstructed mammoth: ruminants rely on a symbiosis with specific microbiota in their stomachs for digestion.

Scientists at the University of Newcastle and University of New South Wales announced in March 2013 that the very recently extinct gastric-brooding frog would be the subject of a cloning attempt to resurrect the species.

Many such "de-extinction" projects are described in the Long Now Foundation›s Revive and Restore Project.

Lifespan

After an eight-year project involving the use of a pioneering cloning technique, Japanese researchers created 25 generations of healthy cloned mice with normal lifespans, demonstrating that clones are not intrinsically shorter-lived than naturally born animals. Other sources have noted that the offspring of clones tend to be healthier than the original clones and indistinguishable from animals produced naturally.

In a detailed study released in 2016 and less detailed studies by others suggest that once cloned animals get past the first month or two of life they are generally healthy. However, early pregnancy loss and neonatal losses are still greater with cloning than natural conception or assisted reproduction (IVF). Current research endeavors are attempting to overcome this problem.

In Popular Culture

In an article in the 8 November 1993 article of *Time*, cloning was portrayed in a negative way, modifying Michelangelo's *Creation of Adam* to depict Adam with five identical hands. *Newsweek›*s 10 March 1997 issue also critiqued the ethics of human cloning, and included a graphic depicting identical babies in beakers.

Cloning is a recurring theme in a wide variety of contemporary science fiction, ranging from action films such as *Jurassic Park* (1993), *The 6th Day* (2000), *Resident Evil* (2002), *Star Wars* (2002) and *The Island* (2005), to comedies such as Woody Allen›s 1973 film *Sleeper*.

Science fiction has used cloning, most commonly and specifically human cloning, due to the fact that it brings up controversial questions of identity. *A Number* is a 2002 play by English playwright Caryl Churchill which addresses the subject of human cloning and identity, especially nature and nurture. The story, set in the near future, is structured around the conflict between a father (Salter) and his sons (Bernard 1, Bernard 2, and Michael Black) – two of whom are clones of the first one. *A Number* was adapted by Caryl Churchill for television, in a co-production between the BBC and HBO Films.

A recurring sub-theme of cloning fiction is the use of clones as a supply of organs for transplantation. The 2005 Kazuo Ishiguro novel *Never Let Me Go* and the 2010 film adaption are set in an alternate history in which cloned humans are created for the sole purpose of providing organ donations to naturally born humans, despite the fact that they are fully sentient and self-aware. The 2005 film *The Island* revolves around a similar plot, with the exception that the clones are unaware of the reason for their existence.

The use of human cloning for military purposes has also been explored in several works. *Star Wars* portrays human cloning in *Clone Wars*.

The exploitation of human clones for dangerous and undesirable work was examined in the 2009 British science fiction film *Moon*. In the futuristic novel *Cloud Atlas* and subsequent film, one of the story lines focuses on a genetically-engineered fabricant clone named Sonmi~451 who is one of millions raised in an artificial "wombtank," destined to serve from birth. She is one of thousands of clones created for manual and emotional labor; Sonmi herself works as a server in a restaurant. She later discovers that the sole source of food for clones, called ‹Soap›, is manufactured from the clones themselves.

Cloning has been used in fiction as a way of recreating historical figures. In the 1976 Ira Levin novel *The Boys from Brazil* and its 1978 film adaptation, Josef Mengele uses cloning to create copies of Adolf Hitler.

In 2012, a Japanese television show named "Bunshin" was created. The story›s main character, Mariko, is a woman studying child welfare in Hokkaido. She grew up always doubtful about the love from her mother, who looked nothing like her and who died nine years before. One day, she finds some of her mother›s belongings at a relative›s house, and heads to Tokyo to seek out the truth behind her birth. She later discovered that she was a clone.

In the 2013 television show *Orphan Black*, cloning is used as a scientific study on the behavioral adaptation of the clones. In a similar vein, the book *The Double* by Nobel Prize winner José Saramago explores the emotional experience of a man who discovers that he is a clone.

GENE MAPPING

Gene mapping describes the methods used to identify the locus of a gene and the distances between genes.

The essence of all genome mapping is to place a collection of molecular markers onto their respective positions on the genome. Molecular markers come in all forms. Genes can be viewed as one special type of genetic markers in the construction of genome maps, and mapped the same way as any other markers.

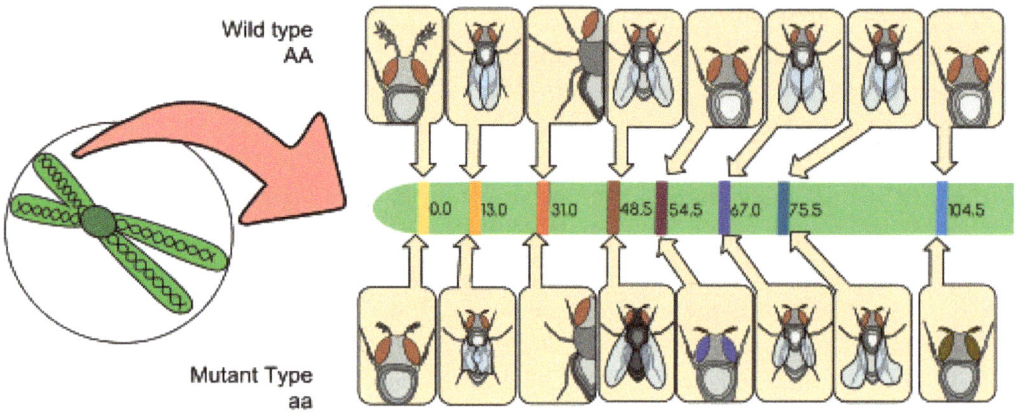

Thomas Hunt Morgan's *Drosophila melanogaster*genetic linkage map. This was the first successful gene mapping work and provides important evidence for the Boveri–Sutton chromosome theory of inheritance. The map shows the relative positions of allelic characteristics on the second Drosophila chromosome. The distance between the genes (map units) are equal to the percentage of crossing-over events that occurs between different alleles.

Gene Mapping VS Physical Mapping

There are two distinctive types of "Maps" used in the field of genome mapping: genetic maps and physical maps. While both maps are a collection of genetic markers and gene location, genetic maps' distances are based on the genetic linkage information, while physical maps use actual physical distances usually measured in number of base pairs. While the physical map could be a more "accurate" representation of the genome, genetic maps often offer insights into the nature of different regions of the chromosome, e.g. the genetic distance to physical distance ratio varies greatly at different genomic regions which reflects different recombination rates, and such rate is often indicative of euchromatic (usually gene-rich) vs heterochromatic (usually gene poor) regions of the genome.

Gene Mapping

Researchers begin a genetic map by collecting samples of blood or tissue from family members that carry a prominent disease or trait and family members that don't. Scientists then isolate DNA from the samples and closely examine it, looking for unique patterns in the DNA of the family members who do carry the disease that the DNA of those who don't carry the disease don't have. These unique molecular patterns in the DNA are referred to as polymorphisms, or markers.

The first steps of building a genetic map are the development of genetic markers and a mapping population. Since the closer the two markers are on the chromosome, the more likely they are to be passed on to the next generation together, therefore the "co-segregation" patterns of all markers can be used to reconstruct their order. With this in mind, the genotypes of each genetic marker are recorded for both parents, and in each individual in the following generations. The quality of the genetic maps is largely dependent upon these two factors: the number of genetic markers on the map and the size of the mapping population. The two factors are interlinked, as a larger mapping population could increase the "resolution" of the map and prevent the map being "saturated".

In gene mapping, any sequence feature that can be faithfully distinguished from the two parents can be used as a genetic marker. Genes, in this regard, are represented by "traits" that can be faithfully distinguished between two parents. Their linkage with other genetic markers are calculated same way as if they are common markers and the actual gene loci are then bracketed in a region between the two nearest neighbouring markers. The entire process is then repeated by looking at more markers which target that region to map the gene neighbourhood to a higher resolution until a specific causative locus can be identified. This process is often referred to as "positional cloning", and it is used extensively in the study of plant species.

Physical Mapping

Since actual base-pair distances are generally hard or impossible to directly measure, physical maps are actually constructed by first shattering the genome into hierarchically smaller pieces. By characterizing each single piece and assembling back together, the overlapping path or "tiling path" of these small fragments would allow researchers to infer physical distances between genomic features. The fragmentation of the genome can be achieved by restriction enzyme cutting or by physically shattering the genome by processes like sonication. Once cut, the DNA fragments are separated by electrophoresis. The resulting pattern of DNA migration (i.e. its genetic fingerprint) is used to identify what stretch of DNA is in the clone. By analyzing the fingerprints, contigs are assembled by automated (FPC) or manual means (Pathfinders) into overlapping DNA stretches. Now a good choice of clones can be made to efficiently sequence the clones to determine the DNA sequence of the organism under study.

In physical mapping, there are no direct ways of marking up a specific gene since the mapping does not include any information that concerns traits and functions. Genetic markers can be linked to a physical map by processes like in situ hybridization. By this approach, physical map contigs can be "anchored" onto a genetic map. The clones used in the physical map contigs can then be sequenced on a local scale to help new genetic marker design and identification of the causative loci.

Macrorestriction is a type of physical mapping wherein the high molecular weight DNA is digested with a restriction enzyme having a low number of restriction sites.

There are alternative ways to determine how DNA in a group of clones overlaps without completely sequencing the clones. Once the map is determined, the clones can be used as a resource to efficiently contain large stretches of the genome. This type of mapping is more accurate than genetic maps.

Genome Sequencing

Genome sequencing is sometimes mistakenly referred to as "genome mapping" by non-biologists. The process of "shotgun sequencing" resembles the process of physical mapping: it shatters the genome into small fragments, characterizes each fragment, then puts them back together (more recent sequencing technologies are drastically different). While the scope, purpose and process are totally different, a genome assembly can be viewed as the "ultimate" form of physical map, in that it provides in a much better way all the information that a traditional physical map can offer.

Use of Gene Mapping

Identification of genes is usually the first step in understanding a genome of a species; mapping of the gene is usually the first step of identification of the gene. Gene mapping is usually the starting point of many important downstream studies.

Disease-Association

The process to identify a genetic element that is responsible for a disease is also referred to as "mapping". If the locus in which the search is performed is already considerably constrained, the search is called the "fine-mapping" of a gene. This information is derived from the investigation of disease-manifestations in large families (Genetic linkage) or from populations-based genetic association studies.

REFERENCES

1. Watson JD (2007). Recombinant DNA: genes and genomes: a short course. San Francisco: W.H. Freeman. ISBN 0-7167-2866-4.
2. Patten CL, Glick BR, Pasternak J (2009). Molecular Biotechnology: Principles and Applications of Recombinant DNA. Washington, D.C: ASM Press. ISBN 978-1-55581-498-4.
3. Brown T (2006). Gene cloning and DNA analysis: an introduction. Cambridge, MA: Blackwell Pub. ISBN 978-1-4051-1121-8.
4. Nathans D, Smith HO (1975). "Restriction endonucleases in the analysis and restructuring of dna molecules". Annual Review of Biochemistry. 44: 273–93. doi:10.1146/annurev.bi.44.070175.001421. PMID 166604.
5. Cohen SN, Chang AC, Boyer HW, Helling RB (Nov 1973). "Construction of biologically functional bacterial plasmids in vitro". Proceedings of the National Academy of Sciences of the United States of America. 70 (11): 3240–4. doi:10.1073/pnas.70.11.3240. PMC 427208 . PMID 4594039.
6. Jackson DA, Symons RH, Berg P (Oct 1972). "Biochemical method for inserting new genetic information into DNA of Simian Virus 40: circular SV40 DNA molecules containing lambda phage genes and the galactose operon of Escherichia coli". Proceedings of the National Academy of Sciences of the United States of

America. 69 (10): 2904–9. doi:10.1073/pnas.69.10.2904. PMC 389671 . PMID 4342968.

7. Shizuya H, Birren B, Kim UJ, Mancino V, Slepak T, Tachiiri Y, Simon M (Sep 1992). "Cloning and stable maintenance of 300-kilobase-pair fragments of human DNA in Escherichia coli using an F-factor-based vector". Proceedings of the National Academy of Sciences of the United States of America. 89 (18): 8794–7. doi:10.1073/pnas.89.18.8794. PMC 50007 . PMID 1528894.
8. Russell DW, Sambrook J (2001). Molecular cloning: a laboratory manual. Cold Spring Harbor, N.Y: Cold Spring Harbor Laboratory. ISBN 978-0-87969-576-7.
9. Higuchi R, Bowman B, Freiberger M, Ryder OA, Wilson AC (1984). "DNA sequences from the quagga, an extinct member of the horse family". Nature. 312 (5991): 282–4. doi:10.1038/312282a0. PMID 6504142.
10. "Torrey Botanical Club: Volumes 42-45". Torreya. Torrey Botanical Club. 42-45: 133. 1942
11. American Association for the Advancement of Science (1903). Science. Moses King. pp. 502–. Retrieved 8 October 2010.
12. de Candolle, A. (1868). Laws of Botanical Nomenclature adopted by the International Botanical Congress held at Paris in August 1867; together with an Historical Introduction and Commentary by Alphonse de Candolle, Translated from the French. translated by H.A. Weddell. London: L. Reeve and Co.
13. "Tasmanian bush could be oldest living organism". Discovery Channel. Archived from the original on 23 July 2006. Retrieved 2008-05-07.
14. "Ibiza's Monster Marine Plant". Ibiza Spotlight. Archived from the original on 26 December 2007. Retrieved 2008-05-07.
15. DeWoody, J.; Rowe, C.A.; Hipkins, V.D.; Mock, K.E. (2008). ""Pando" Lives: Molecular Genetic Evidence of a Giant Aspen Clone in Central Utah". Western North American Naturalist. 68 (4): 493–497. doi:10.3398/1527-0904-68.4.493.
16. Mock, K.E. , Rowe, C.A. , Hooten, M.B. , Dewoody, J. , Hipkins, V.D. (2008). "Blackwell Publishing Ltd Clonal dynamics in western North American aspen (Populus tremuloides)". U.S. Department of Agriculture, Oxford, UK : Blackwell Publishing Ltd. p. 17. Retrieved 2013-12-05.
17. Peter J. Russel (2005). iGenetics: A Molecular Approach. San Francisco, California, United States of America: Pearson Education. ISBN 0-8053-4665-1.
18. McFarland, Douglas (2000). "Preparation of pure cell cultures by cloning". Methods in Cell Science. 22 (1): 63–66. doi:10.1023/A:1009838416621. PMID 10650336.
19. Gil, Gideon (2008-01-17). "California biotech says it cloned a human embryo, but no stem cells produced". Boston Globe.
20. Halim, N. (September 2002). "Extensive new study shows abnormalities in cloned animals". Massachusetts institute of technology. Retrieved October 2011. Check date values in: |access-date= (help)

21. Plus, M. (2011). "Fetal development". Nlm.nih.gov. Retrieved October 2011. Check date values in: |access-date= (help)
22. Latham, K. E. (2005). "Early and delayed aspects of nuclear reprogramming during cloning" (PDF). Biology of the Cell. pp. 97, 119–132.
23. Campbell KH, McWhir J, Ritchie WA, Wilmut I (March 1996). "Sheep cloned by nuclear transfer from a cultured cell line". Nature. 380 (6569): 64–6. doi:10.1038/380064a0. PMID 8598906.
24. Shukman, David (14 January 2014) China cloning on an 'industrial scale' BBC News Science and Environment, Retrieved 10 April 2014
25. Zastrow, Mark (8 February 2016). "Inside the cloning factory that creates 500 new animals a day". New Scientist. Retrieved 23 February 2016.
26. "Asexual Propagation". Aggie-horticulture.tamu.edu. Retrieved 2010-08-04.
27. Sagers, Larry A. (2 March 2009) Proliferate favorite trees by grafting, cloningUtah Stae University, Deseret News (Salt Lake City) Retrieved 21 February 2014
28. Perrier, X.; De Langhe, E.; Donohue, M.; Lentfer, C.; Vrydaghs, L.; Bakry, F.; Carreel, F.; Hippolyte, I.; Horry, J. -P.; Jenny, C.; Lebot, V.; Risterucci, A. -M.; Tomekpe, K.; Doutrelepont, H.; Ball, T.; Manwaring, J.; De Maret, P.; Denham, T. (2011). "Multidisciplinary perspectives on banana (Musa spp.) domestication"(PDF). Proceedings of the National Academy of Sciences. 108 (28): 11311–11318. doi:10.1073/pnas.1102001108.
29. Castagnone-Sereno P, et al. Diversity and evolution of root-knot nematodes, genus Meloidogyne: new insights from the genomic era Annu Rev Phytopathol. 2013;51:203-20
30. Shubin, Neil (24 February 2008) Birds Do It. Bees Do It. Dragons Don't Need To New York Times, Retrieved 21 February 2014
31. Explanation of the Spemann-Mangold experiment from a Nature Reviews article http://www.nature.com/nrm/journal/v7/n4/box/nrm1855_BX1.html
32. "Cloning Fact Sheet". Human Genome Project Information. Archived from the original on 2 May 2013. Retrieved 25 October 2011.
33. Illmensee K, Levanduski M, Vidali A, Husami N, Goudas VT (February 2009). "Human embryo twinning with applications in reproductive medicine". Fertil. Steril. 93 (2): 423–7. doi:10.1016/j.fertnstert.2008.12.098. PMID 19217091.
34. Rantala, Milgram, M., Arthur (1999). Cloning: For and Against. Chicago, Illinois: Carus Publishing Company. p. 1. ISBN 0-8126-9375-2.
35. Swedin, Eric. "Cloning". CredoReference. Retrieved 23 September 2013.
36. Lassen, J.; Gjerris, M.; Sandøe, P. (2005). "After Dolly—Ethical limits to the use of biotechnology on farm animals". Theriogenology. 65: 992–1004. doi:10.1016/j.theriogenology.2005.09.012.
37. Swedin, Eric. "Cloning". CredoReference. Science in the Contemporary World. Retrieved 23 September 2013.
38. TV documentary Visions Of The Future part 2 shows this process, explores the

social implicatins of cloning and contains footage of monoculture in livestock

39. Gurdon J (Apr 1962). "Adult frogs derived from the nuclei of single somatic cells". Dev Biol. 4: 256–73. doi:10.1016/0012-1606(62)90043-x. PMID 13903027.
40. Wakayama S1, Kohda T, Obokata H, Tokoro M, Li C, Terashita Y, Mizutani E, Nguyen VT, Kishigami S, Ishino F, Wakayama T (7 March 2013). "Successful serial recloning in the mouse over multiple generations". Cell Stem Cell. 12 (3): 293–7. doi:10.1016/j.stem.2013.01.005. PMID 23472871.
41. BBC. 22 February 2008. BBC On This Day: 1997: Dolly the sheep is cloned
42. "ThinkQuest". Retrieved 3 May 2015.
43. "Robert W. Briggs". National Academies Press. Retrieved 1 December 2012.
44. "Bloodlines timeline". PBS.org.
45. "Кто изобрел клонирование?". Archived from the original on 2004-12-23. (Russian)
46. "Gene Genie | BBC World Service". Bbc.co.uk. 2000-05-01. Retrieved 2010-08-04.
47. McLaren A (2000). "Cloning: pathways to a pluripotent future". Science. 288 (5472): 1775–80. doi:10.1126/science.288.5472.1775. PMID 10877698.
48. CNN. Researchers clone monkey by splitting embryo 2000-01-13. Retrieved 2008-08-05.
49. Dean Irvine (2007-11-19). "You, again: Are we getting closer to cloning humans? - CNN.com". Edition.cnn.com. Retrieved 2010-08-04.
50. Grisham, Julie (April 2000). "Pigs cloned for first time". Nature Biotechnology. 18 (4): 365. doi:10.1038/74335.
51. Shukman, David (14 January 2014) China cloning on an ‹industrial scale› BBC News Science and Environment, Retrieved 14 January 2014
52. "First cloned endangered species dies 2 days after birth". CNN. 12 January 2001. Retrieved 30 April 2010.
53. Camacho, Keite. Embrapa clona raça de boi ameaçada de extinção. Agência Brasil. 2005-05-20. (Portuguese) Retrieved 2008-08-05
54. "Americas | Pet kitten cloned for Christmas". BBC News. 2004-12-23. Retrieved 2010-08-04.
55. "Rat called Ralph is latest clone". BBC News. 25 September 2003. Retrieved 30 April 2010.
56. Associated Press 25 August 2009 (2009-08-25). "Gordon Woods dies at 57; Veterinary scientist helped create first cloned mule". latimes.com. Retrieved 2010-08-04.
57. "World›s first cloned horse is born - 06 August 2003". New Scientist. Retrieved 2010-08-04.
58. "First Dog Clone". News.nationalgeographic.com. Retrieved 2010-08-04.
59. (1 September 2009) World›s first cloned wolf dies Phys.Org, Retrieved 9 April

2015

60. Sinha, Kounteya (2009-02-13). “India clones world›s first buffalo”. The Times of India. Retrieved 2010-08-04.
61. Note: The Pyrenean ibex is an extinct sub-species; the broader species, the Spanish ibex, is thriving. Peter Maas. Pyrenean Ibex - Capra pyrenaica pyrenaicaat The Sixth Extinction]. Last updated 15 April 2012.
62. Extinct ibex is resurrected by cloning, The Daily Telegraph, 31 January 2009
63. Spencer, Richard (14 April 2009). “World›s first cloned camel unveiled in Dubai”. London: Telegraph.co.uk. Retrieved 15 April 2009.
64. Ishfaq-ul-Hassan (15 March 2012). “India gets its second cloned animal Noorie, a pashmina goat”. Kashmir, India: DNA.
65. Hickman, L. “Scientists clone extinct frog – Jurassic Park here we come?”. The Guardian. Retrieved July 10, 2016.
66. Kfoury, C (July 2007). “Therapeutic cloning: promises and issues”. Mcgill J Med. 10: 112–20. PMC 2323472 . PMID 18523539.
67. “Cloning Fact Sheet”. U.S. Department of Energy Genome Program. 2009-05-11. Archived from the original on 2013-05-02.
68. de Grey, Aubrey; Rae, Michael (September 2007). Ending Aging: The Rejuvenation Breakthroughs that Could Reverse Human Aging in Our Lifetime. New York, NY: St. Martin›s Press, 416 pp. ISBN 0-312-36706-6.
69. Staff, Times Higher Education. 10 August 2001 In the news: Antinori and Zavos
70. “AAAS Statement on Human Cloning”.
71. McGee, G. (October 2011). “Primer on Ethics and Human Cloning”. American Institute of Biological Sciences.
72. “Universal Declaration on the Human Genome and Human Rights”. UNESCO. 1997-11-11. Retrieved 2008-02-27.
73. McGee, Glenn (2000). ‹The Perfect Baby: Parenthood in the New World of Cloning and Genetics.› Lanham: Rowman & Littlefield.
74. Havstad JC (2010). “Human reproductive cloning: a conflict of liberties.”. Bioethics. 24 (2): 71–7. doi:10.1111/j.1467-8519.2008.00692.x. PMID 19076121.
75. Bob Sullivan, Technology correspondent for MSNBC. November 262003 Religions reveal little consensus on cloning - Health - Special Reports - Beyond Dolly: Human Cloning
76. William Sims Bainbridge, Ph.D. Religious Opposition to Cloning Journal of Evolution and Technology - Vol. 13 - October 2003
77. Staff, Humane Society Cloning
78. Sean Poulter for the Daily Mail. 26 November 2010. Clone farming would introduce cruelty on a massive scale, say animal welfare groups
79. “Effect of calf death loss on cloned cattle herd derived from somatic cell nuclear transfer: clones with congenital defects would be removed by the death loss.”.

Animal Science Journal. 84: 631–8. Sep 2013. doi:10.1111/asj.12087. PMID 23829575.

80. "FDA says cloned animals are OK to eat". NBCNews.com. Associated Press. 28 December 2006.
81. "An HSUS Report: Welfare Issues with Genetic Engineering and Cloning of Farm Animals" (PDF). Humane Society of the United States.
82. "Not Ready for Prime Time: FDA›s Flawed Approach to Assessing the Safety of Food from Animal Clones" (PDF). Center for Food Safety. March 2007.(PDF)
83. Hansen, Michael (2007-04-27). "Comments of Consumers Union to US Food and Drug Administration on Docket No. 2003N-0573, Draft Animal Cloning Risk Assessment" (PDF). Consumers Union.
84. Holt, W. V., Pickard, A. R., & Prather, R. S. (2004) Wildlife conservation and reproductive cloning. Reproduction, 126.
85. Ehrenfeld, David (2006). "Transgenics and Vertebrate Cloning as Tools for Species Conservation". Conservation Biology. 20 (3): 723–732. doi:10.1111/j.1523-1739.2006.00399.x. PMID 16909565.
86. Ono T, Li C, Mizutani E, Terashita Y, Yamagata K, Wakayama T (Dec 2010). "Inhibition of class IIb histone deacetylase significantly improves cloning efficiency in mice". Biol Reprod. 83 (6): 929–37. doi:10.1095/biolreprod.110.085282. PMID 20686182.
87. Shukman, David (14 January 2014) China cloning on an ‹industrial scale› BBC News Science and Environment, Retrieved 27 February 2016
88. Baer, Drake (8 September 2015). "This Korean lab has nearly perfected dog cloning, and that›s just the start". Tech Insider, Innovation. Retrieved 27 February 2016.
89. Ferris Jabr for Scientific American. 11 March 2013. Will cloning ever saved endangered species?
90. Heidi B. Perlman (2000-10-08). "Scientists Close on Extinct Cloning". The Washington Post. Associated Press.
91. Pence, Gregory E. (2005). Cloning After Dolly: Who›s Still Afraid?. Rowman & Littlefield. ISBN 0-7425-3408-1.
92. Holloway, Grant (2002-05-28). "Cloning to revive extinct species". CNN.com.
93. Gray, Richard; Dobson, Roger (31 January 2009). "Extinct ibex is resurrected by cloning". The Telegraph. London. Retrieved 2009-02-01.
94. "Scientists ‹to clone mammoth›". BBC News. 2003-08-18.
95. "BBC News". Bbc.co.uk. 2011-12-07. Retrieved 2012-08-19.
96. "Когда вернутся мамонты» ("When the Mammoths Return"), February 5, 2015 (retrieved September 6, 2015)
97. Yong, Ed (2013-03-15). "Resurrecting the Extinct Frog with a Stomach for a Womb". National Geographic. Retrieved 2013-03-15.
98. "Long Now Foundation, Revive and Restore Project".

99. "Generations of Cloned Mice With Normal Lifespans Created: 25th Generation and Counting". Science Daily. 7 March 2013. Retrieved 8 March 2013.
100. Carey, Nessa (2012). The Epigenetics Revolution. London, UK: Icon Books Ltd. pp. 149–150. ISBN 978-184831-347-7.
101. planktonrules (17 December 1973). "Sleeper (1973)". IMDb. Retrieved 3 May 2015.
102. Hopkins, Patrick. "How Popular media represent cloning as an ethical problem". 28. The Hastings Center: 6–13. doi:10.2307/3527566. JSTOR 3527566.
103. "Yvonne A. De La Cruz ‹›Science Fiction Storytelling and Identity: Seeing the Human Through Android Eyes››" (PDF). Retrieved 2012-08-19.
104. "Uma Thurman, Rhys Ifans and Tom Wilkinson star in two plays for BBC Two"(Press release). BBC. 2008-06-19. Retrieved 2008-09-09.
105. espenshade55 (11 February 2011). "Never Let Me Go (2010)". IMDb. Retrieved 3 May 2015.
106. "The Island (2005)". IMDb. 22 July 2005. Retrieved 3 May 2015.
107. compel_bast (15 August 2008). "Star Wars: The Clone Wars (2008)". IMDb. Retrieved 3 May 2015.
108. larry-411 (17 July 2009). "Moon (2009)". IMDb. Retrieved 3 May 2015.
109. Could the Human Clones of ‹Cloud Atlas› Be in Our Future?
110. Christian Lee Pyle (CLPyle) (12 October 1978). "The Boys from Brazil (1978)". IMDb. Retrieved 3 May 2015.
111. "Review of Bunshin".
112. foreverbounds. "Orphan Black (TV Series 2013–)". IMDb. Retrieved 3 May 2015.
113. Banville, John (2004-10-10). "›The Double›: The Tears of a Clone". The New York Times. Retrieved 14 January 2015.
114. Lederberg J (Feb 1994). "The transformation of genetics by DNA: an anniversary celebration of Avery, MacLeod and McCarty (1944)". Genetics. 136 (2): 423–6. PMC 1205797 . PMID 8150273.
115. Wirth R, Friesenegger A, Fiedler S (Mar 1989). "Transformation of various species of gram-negative bacteria belonging to 11 different genera by electroporation". Molecular & General Genetics. 216 (1): 175–7. doi:10.1007/BF00332248. PMID 2659971.
116. Oldenburg J, Dolan G, Lemm G (Jan 2009). "Haemophilia care then, now and in the future". Haemophilia. 15 Suppl 1: 2–7. doi:10.1111/j.1365-2516.2008.01946.x. PMID 19125934.
117. The MJ (Nov 1989). "Human insulin: DNA technology›s first drug". American Journal of Hospital Pharmacy. 46 (11 Suppl 2): S9–11. PMID 2690608.
118. Lewandowski C, Barsan W (Feb 2001). "Treatment of acute ischemic stroke". Annals of Emergency Medicine. 37 (2): 202–16. doi:10.1067/mem.2001.111573. PMID 11174240.

119. Chang MH, Chen CJ, Lai MS, Hsu HM, Wu TC, Kong MS, Liang DC, Shau WY, Chen DS (Jun 1997). "Universal hepatitis B vaccination in Taiwan and the incidence of hepatocellular carcinoma in children. Taiwan Childhood Hepatoma Study Group". The New England Journal of Medicine. 336 (26): 1855–9. doi:10.1056/NEJM199706263362602. PMID 9197213.

120. August JT (1997). Gene Therapy. 40. Academic Press. p. 508. ISBN 978-0-08-058132-3.

121. Pfeifer A, Verma IM (2001). "Gene therapy: promises and problems". Annual Review of Genomics and Human Genetics. 2: 177–211. doi:10.1146/annurev.genom.2.1.177. PMID 11701648.

CHAPTER 6

Restriction Mapping

Chapter Contents

Restriction mapping is the process of obtaining structural information on a piece of DNA by the use of restriction enzymes. It is used to map an unknown segment of DNA by breaking it into pieces and identifying the locations of the breakpoints.

INTRODUCTION

A restriction map is a map of known restriction sites within a sequence of DNA. Restriction mapping requires the use of restriction enzymes. In molecular biology, restriction maps are used as a reference to engineer plasmids or other relatively short pieces of DNA, and sometimes for longer genomic DNA. There are other ways of mapping features on DNA for longer length DNA molecules, such as mapping by transduction.

One approach in constructing a restriction map of a DNA molecule is to sequence the whole molecule and to run the sequence through a computer program that will find the recognition sites that are present for every restriction enzyme known.

Before sequencing was automated, it would have been prohibitively expensive to sequence an entire DNA strand. To find the relative positions of restriction sites on a plasmid, a technique involving single and double restriction digests is used. Based on the sizes of the resultant DNA fragments the positions of the sites can be inferred. Restriction mapping is a very useful technique when used for determining the orientation of an insert in a cloning vector, by mapping the position of an off-center restriction site in the insert.

METHOD

The experimental procedure first requires an aliquot of purified plasmid DNA for each digest to be run. Digestion is then performed with each enzyme(s) chosen. The resulting samples are subsequently run on an electrophoresis gel, typically on agarose gel.

The first step following the completion of electrophoresis is to add up the sizes of the fragments in each lane. The sum of the individual fragments should equal the size of the original fragment, and each digest's fragments should also sum up to be the same size as each other. If fragment sizes do not properly add up, there are two likely problems. In one case, some of the smaller fragments may have run off the end of the gel. This frequently occurs if the gel is run too long. A second possible source of error is that the gel was not dense enough and therefore was unable to resolve fragments close in size. This leads to a lack of separation of fragments which were close in size. If all of the digests produce fragments that add up one may infer the position of the REN (restriction endonuclease) sites by placing them in spots on the original DNA fragment that would satisfy the fragment sizes produced by all three digests.

EXAMPLE

For example the most common application of restriction mapping is presented: Determining the orientation of a cloned insert. This method requires that restriction maps of the cloning vector and the insert are already available.

If you know of a restriction site placed towards one end of the insert you can determine the orientation by observing the size of the fragments in the gel. Often the orientation of inserts is important and this technique is used to screen for the correct orientation.

In this example the orientation of an insert cloned with EcoRI will be found.

Digests

- 1: EcoRI
- 2: HindIII
- 3: EcoRI + HindIII (optional and not discussed)

Resultant Fragments: approximate sizes

- 1: 3 kb, 5 kb
- 2: 2 kb, 6 kb
- 3: 2 kb, 1 kb, 5 kb,

Hypothetical Multiple Cloning Site of Vector

5'-----HindIII-EcoRI----3'

Discussion

The EcoRI digest excises the insert yielding fragments of 3 kb and 5 kb. These are the sizes of the insert and vector backbone respectively. This is expected since the size of both the insert and vector are known beforehand. The presence of the insert is confirmed.

There is a known HindIII site off-center in the 3 kb insert. It is 2 kb away from one end (end A), and 1 kb away from the other end (end B). The HindIII digest of your clone yields fragments of 2 kb and 6 kb. The 2 kb fragment is exclusively the insert sequence and the 6 kb fragment is 1 kb of insert sequence attached to 5 kb of vector sequence. This means that the insert was cloned in an A to B orientation as opposed to B to A which would yield fragments of 7 kb and 1kb.

Resultant Map

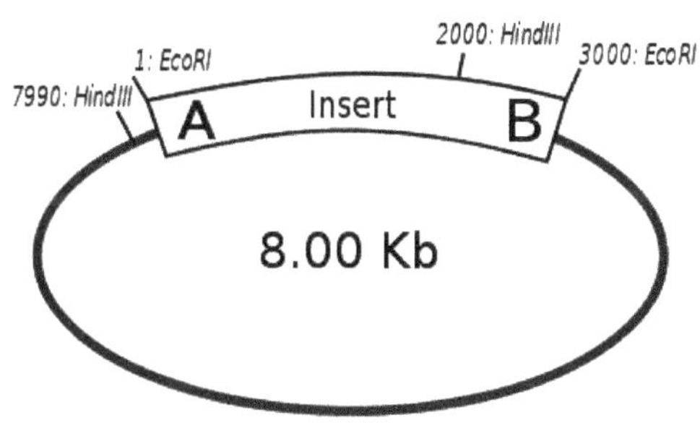

Resultant map

Appendix: Related Methods

Rapid Denaturation and Renaturation of a crude DNA preparation by alkaline lysis of the cells and subsequent neutralization

In this technique the cells are lysed in alkaline conditions. The DNA in the mixture is denatured (strands separated) by disrupting the hydrogen bonds between the two strands. The large genomic DNA is subject to tangling and staying denatured when the pH is lowered during the neutralization. In other words, the strands come back together in a disordered fashion, basepairing randomly. The circular supercoiled plasmids' strands will stay relatively closely aligned and will renature correctly. Therefore, the genomic DNA will form an insoluble aggregate and the supercoiled plasmids will be left in solution. This can be followed by phenol extraction to remove proteins and other molecules. Then the DNA can be subjected to ethanol precipitation to concentrate the sample.

REFERENCES

1. Bitner, R; Kuempel, Peter (February 1982). "P1 Transduction Mapping of the trg Locus in rac+ and rac Strains of Escherichia coli K-12" (PDF). Journal of Bacteriology. 149 (2): 529–533.
2. Dale, J; von Schantz, M; Greenspan, D (2003). From Genes to Genomes. West Sussex: John Wiley & Sons Ltd.

CHAPTER 7

Molecular Modelling

Chapter Contents

Molecular modelling is a collection of (computer based) techniques for deriving, representing and manipulating the structures and reactions of molecules, and those properties that are dependent on these three dimensional structures. The emphasis will be on helping to develop a feel for the correct "tool" to use in the context of a typical problem in structure, activity or reactivity, by describing the limitations and strengths of each method. The techniques described include:

- Molecular visualisation
- Methods for Equilibrium and transition state geometry location
- Introduction to Molecular mechanics methods
- Applications of Semi-empirical, *ab initio* and density functional molecular orbital methods
- Methods for topological analysis of wavefunctions

INTRODUCTION

Molecular modelling encompasses all methods, theoretical and computational, used to model or mimic the behaviour of molecules. The methods are used in the fields of computational chemistry, drug design, computational biology and materials science to study molecular systems ranging from small chemical systems to large biological molecules and material assemblies. The simplest calculations can be performed by hand, but inevitably computers are required to perform molecular modelling of any reasonably sized system. The common feature of molecular modelling methods is the atomistic level description of the molecular systems. This may include treating atoms as the smallest individual unit (a molecular mechanics approach), or explicitly modelling electrons of each atom (a quantum chemistry approach).

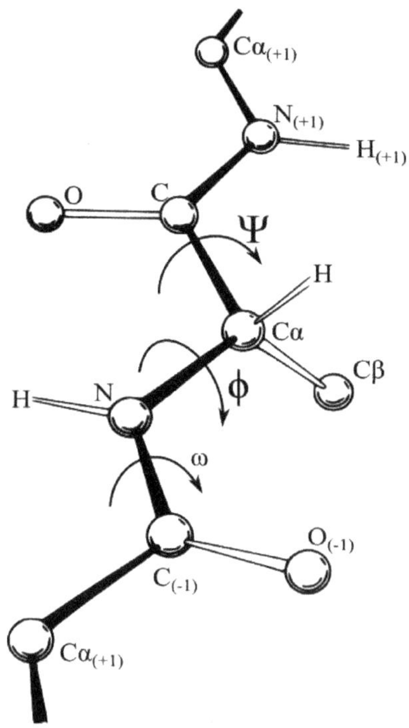

The backbone dihedral angles are included in the molecular model of a protein.

Molecular Mechanics

Molecular mechanics is one aspect of molecular modelling, as it refers to the use of classical mechanics (Newtonian mechanics) to describe the physical basis behind the models. Molecular models typically describe atoms (nucleus and electrons collectively) as point charges with an associated mass. The interactions between neighbouring atoms are described by spring-like interactions (representing chemical bonds) and Van der Waals forces. The Lennard-Jones potential is commonly used to describe the later. The electrostatic interactions are computed based on Coulomb›s law. Atoms are assigned coordinates in Cartesian space or in internal coordinates, and can also be assigned velocities in dynamical simulations. The atomic velocities are related to the temperature of the system, a macroscopic quantity. The collective mathematical expression is termed a potential function and is related to the system internal energy (U), a thermodynamic quantity equal to the sum of potential and kinetic energies. Methods which minimize the potential energy are termed energy minimization methods (e.g., steepest descent and conjugate gradient), while methods that model the behaviour of the system with propagation of time are termed molecular dynamics.

$$E = E_{\text{bonds}} + E_{\text{angle}} + E_{\text{dihedral}} + E_{\text{non-bonded}}$$

$$E_{\text{non-bonded}} = E_{\text{electrostatic}} + E_{\text{van der Waals}}$$

This function, referred to as a potential function, computes the molecular potential energy as a sum of energy terms that describe the deviation of bond lengths, bond angles and torsion angles away from equilibrium values, plus terms for non-bonded pairs of atoms describing van der Waals and electrostatic interactions. The set of parameters consisting of equilibrium bond lengths, bond angles, partial charge values, force constants and van der Waals parameters are collectively termed a force field. Different implementations of molecular mechanics use different mathematical expressions and different parameters for the potential function. The common force fields in use today have been developed by using high level quantum calculations and/or fitting to experimental data. The method, termed energy minimization, is used to find positions of zero gradient for all atoms, in other words, a local energy minimum. Lower energy states are more stable and are commonly investigated because of their role in chemical and biological processes. A molecular dynamics simulation, on the other hand, computes the behaviour of a system as a function of time. It involves solving Newton's laws of motion, principally the second law, $\mathbf{F} = m\mathbf{a}$. Integration of Newton's laws of motion, using different integration algorithms, leads to atomic trajectories in space and time. The force on an atom is defined as the negative gradient of the potential energy function. The energy minimization method is useful to obtain a static picture for comparing between states of similar systems, while molecular dynamics provides information about the dynamic processes with the intrinsic inclusion of temperature effects.

VARIABLES

Molecules can be modelled either in vacuum, or in the presence of a solvent such as water. Simulations of systems in vacuum are referred to as *gas-phase* simulations, while those that include the presence of solvent molecules are referred to as *explicit solvent* simulations. In another type of simulation, the effect of solvent is estimated using an empirical mathematical expression; these are termed *implicit solvation* simulations.

Coordinate Representations

Most force fields are distance-dependent, making the most convenient expression for these Cartesian coordinates. Yet the comparatively rigid nature of bonds which occur between specific atoms, and in essence, defines what is meant by the designation *molecule*, make an internal coordinate system the most logical representation. In some fields the IC representation (bond length, angle between bonds, and twist angle of the bond as shown in the figure) is termed the Z-matrix or torsion angle representation. Unfortunately, continuous motions in Cartesian space often require discontinuous angular branches in internal coordinates, making it relatively hard to work with force fields in the internal coordinate representation, and conversely a simple displacement of an atom in Cartesian space may not be a straight line trajectory due to the prohibitions of the interconnected bonds. Thus, it is very common for computational optimizing programs to flip back and forth between representations during their iterations. This can dominate the calculation time of the potential itself and in long chain molecules introduce cumulative numerical inaccuracy. While all conversion algorithms produce mathematically identical results, they differ in speed and numerical accuracy. Currently, the fastest and most accurate torsion to Cartesian conversion is the Natural Extension Reference Frame (NERF) method.

APPLICATIONS

Molecular modelling methods are now used routinely to investigate the structure, dynamics, surface properties, and thermodynamics of inorganic, biological, and polymeric systems. The types of biological activity that have been investigated using molecular modelling include protein folding, enzyme catalysis, protein stability, conformational changes associated with biomolecular function, and molecular recognition of proteins, DNA, and membrane complexes.

REFERENCES

1. Parsons, J., Holmes, J. B., Rojas, J. M., Tsai, J., Strauss, C. E., Practical conversion from torsion space to cartesian space for in silico protein synthesis. J Comput Chem 26 (2005), 1063-1068.
2. M. P. Allen, D. J. Tildesley, Computer simulation of liquids, 1989, Oxford University Press, ISBN 0-19-855645-4.
3. D. Frenkel, B. Smit, Understanding Molecular Simulation: From Algorithms to Applications, 1996, ISBN 0-12-267370-0
4. D. C. Rapaport, The Art of Molecular Dynamics Simulation, 2004, ISBN 0-521-82568-7
5. R. J. Sadus, Molecular Simulation of Fluids: Theory, Algorithms and Object-Orientation, 2002, ISBN 0-444-51082-6
6. K.I.Ramachandran, G Deepa and Krishnan Namboori. P.K. Computational Chemistry and Molecular Modeling Principles and Applications 2008 ISBN 978-3-540-77302-3 Springer-Verlag GmbH

CHAPTER 8

Genomic Libraries

Chapter Contents

- **Introduction**
- **Genomic Library Construction**
- **Types of Vectors**
- **How to Select a Vector**

A genomic library is a collection of bacteria which have been genetically engineered to hold the entire DNA of an organism. The size of the library varies, depending on how the DNA is stored in the bacteria, and the length of the genome of the organism. Genomic libraries are used in genetic research all over the world in various lab facilities. Companies which manufacture genomic libraries can provide them by special order to researchers.

INTRODUCTION

A genomic library is a collection of the total genomic DNA from a single organism. The DNA is stored in a population of identical vectors, each containing a different insert of DNA. In order to construct a genomic library, the organism's DNA is extracted from cells and then digested with a restriction enzyme to cut the DNA into fragments of a specific size. The fragments are then inserted into the vector using DNA ligase. Next, the vector DNA can be taken up by a host organism - commonly a population of Escherichia coli or yeast - with each cell containing only one vector molecule. Using a host cell to carry the vector allows for easy amplification and retrieval of specific clones from the library for analysis.

There are several kinds of vectors available with various insert capacities. Generally, libraries made from organisms with larger genomes require vectors featuring larger inserts, thereby fewer vector molecules are needed to make the library. Researchers can choose a vector also considering the ideal insert size to find a desired number of clones necessary for full genome coverage.

Genomic libraries are commonly used for sequencing applications. They have played an important role in the whole genome sequencing of several organisms, including the human genome and several model organisms.

The first DNA-based genome ever fully sequenced was achieved by two-time Nobel Prize winner, Frederick Sanger, in 1977. Sanger and his team of scientists created a library of the bacteriophage, phi X 174, for use in DNA sequencing. The importance of this success contributed to the ever-increasing demand for sequencing genomes to research gene therapy. Teams are now able to catalog polymorphisms in genomes and investigate those candidate genes contributing to maladies such as Parkinson's disease, Alzheimer's disease, multiple sclerosis, rheumatoid arthritis, and Type 1 diabetes. These are due to the advance of genome-wide association studies from the ability to create and sequence genomic libraries. Prior, linkage and candidate-gene studies were some of the only approaches.

GENOMIC LIBRARY CONSTRUCTION

Construction of a genomic library involves creating many recombinant DNA molecules. An organism's genomic DNA is extracted and then digested with a restriction enzyme. For organisms with very small genomes *(~10 kb)*, the digested fragments can be separated by gel electrophoresis. The separated fragments can then be excised and cloned into the vector separately. However, when a large genome is digested with a restriction enzyme, there are far too many fragments to excise individually. The entire set of fragments must be cloned together with the vector, and separation of clones can occur after. In either case, the fragments are ligated into a vector that has been digested with the same restriction enzyme. The vector containing the inserted fragments of genomic DNA can then be introduced into a host organism.

Below are the steps for creating a genomic library from a large genome.

- Extract and purify DNA.
- Digest the DNA with a restriction enzyme. This creates fragments that are similar in size, each containing one or more genes.
- Insert the fragments of DNA into vectors that were cut with the same restriction enzyme. Use the enzyme DNA ligase to seal the DNA fragments into the vector. This creates a large pool of recombinant molecules.
- These recombinant molecules are taken up by a host bacterium by transformation, creating a DNA library.

Below is a diagram of the above outlined steps.

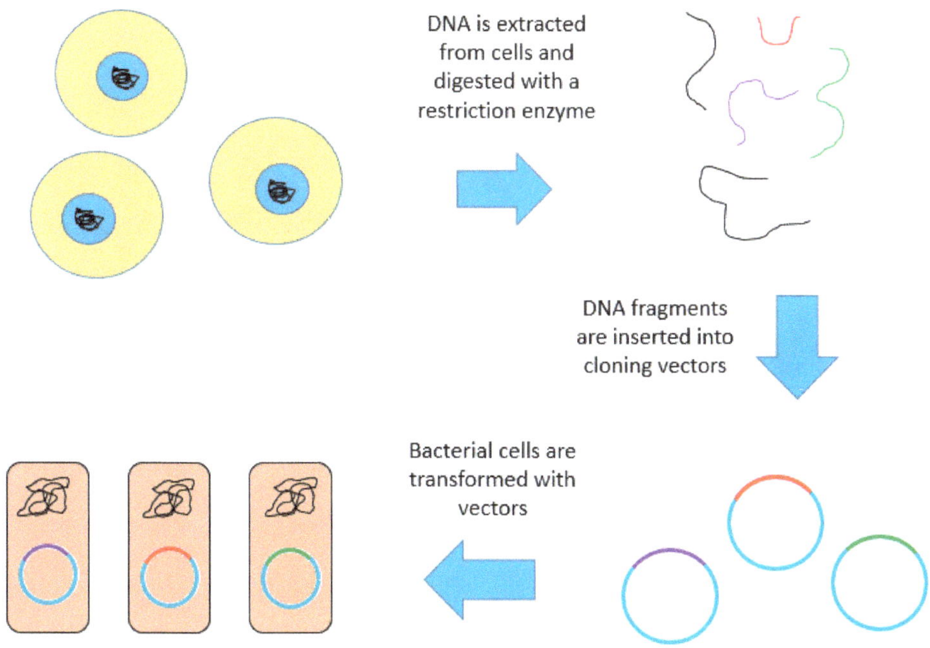

Genomic Library Construction

Determining Titer of Library

After a genomic library is constructed with a viral vector, such as lambda phage, the titer of the library can be determined. Calculating the titer allows researchers to approximate how many infectious viral particles were successfully created in the library. To do this, dilutions of the library are used to transfect cultures of E. coli of known concentrations. The cultures are then plated on agar plates and incubated overnight. The number of viral plaques are counted and can be used to calculate the total number of infectious viral particles in the library. Most viral vectors also carry a marker that allows clones containing an insert to be distinguished from those that do not have an insert. This allows researchers to also determine the percentage of infectious viral particles actually carrying a fragment of the library.

A similar method can be used to titer genomic libraries made with non-viral vectors, such as plasmids and BACs. A test ligation of the library can be used to transform E. coli. The transformation is then spread on agar plates and incubated overnight. The titer of the transformation is determined by counting the number of colonies present on the plates. These vectors generally have a selectable marker allowing the differentiation of clones containing an insert from those that do not. By doing this test, researchers can also determine the efficiency of the ligation and make adjustments as needed to ensure they get the desired number of clones for the library.

Screening Library

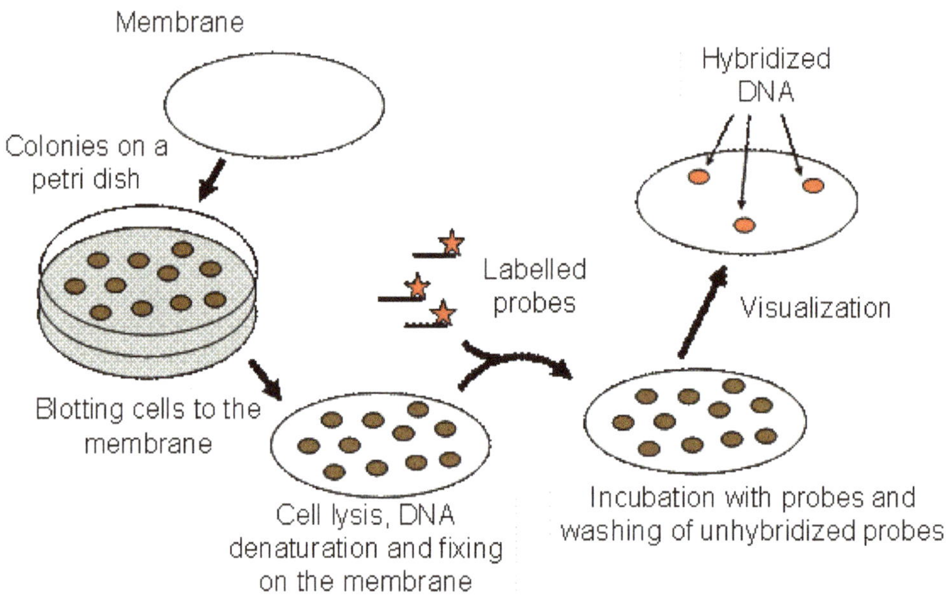

Colony Blot Hybridization

In order to isolate clones that contain regions of interest from a library, the library must first be screened. One method of screening is hybridization. Each transformed host cell of a library will contain only one vector with one insert of DNA. The whole library can be plated onto a filter over media. The filter and colonies are prepared for hybridization and then labeled with a probe. The target DNA- insert of interest- can be identified by detection such as autoradiography because of the hybridization with the probe as seen below.

Another method of screening is with polymerase chain reaction (PCR). Some libraries are stored as pools of clones and screening by PCR is an efficient way to identify pools containing specific clones.

TYPES OF VECTORS

Genome size varies among different organisms and the cloning vector must be selected accordingly. For a large genome, a vector with a large capacity should be chosen so that a relatively small number of clones are sufficient for coverage of the entire genome. However, it is often more difficult to characterize an insert contained in a higher capacity vector.

Below is a table of several kinds of vectors commonly used for genomic libraries and the insert size that each generally holds.

Vector type	Insert size (thousands of bases)
Plasmids	up to 15
Phage lambda (λ)	up to 25
Cosmids	up to 45
Bacteriophage P1	70 to 100
P1 artificial chromosomes (PACs)	130 to 150
Bacterial artificial chromosomes (BACs)	120 to 300
Yeast artificial chromosomes (YACs)	250 to 2000

Plasmids

A plasmid is a double stranded circular DNA molecule commonly used for molecular cloning. Plasmids are generally 2 to 4 kilobase-pairs (kb) in length and are capable of carrying inserts up to 15kb. Plasmids contain an origin of replication allowing them to replicate inside a bacterium independently of the host chromosome. Plasmids commonly carry a gene for antibiotic resistance that allows for the selection of bacterial cells containing the plasmid. Many plasmids also carry a reporter gene that allows researchers to distinguish clones containing an insert from those that do not.

Phage Lambda (λ)

Phage λ is a double-stranded DNA virus that infects *E. coli*. The λ chromosome is 48.5kb long and can carry inserts up to 25kb. These inserts replace non-essential viral sequences in the λ chromosome, while the genes required for formation of viral particles and infection remain intact. The insert DNA is replicated with the viral DNA; thus, together they are packaged into viral particles. These particles are very efficient at infection and multiplication leading to a higher production of the recombinant λ chromosomes. However, due to the smaller insert size, libraries made with λ phage may require many clones for full genome coverage.

Cosmids

Cosmid vectors are plasmids that contain a small region of bacteriophage λ DNA called the cos sequence. This sequence allows the cosmid to be packaged into bacteriophage λ particles. These particles- containing a linearized cosmid- are introduced into the host cell by transduction. Once inside the host, the cosmids circularize with the aid of the host›s DNA ligase and then function as plasmids. Cosmids are capable of carrying inserts up to 45kb in size.

Bacteriophage P1 Vectors

Bacteriophage P1 vectors can hold inserts 70 – 100kb in size. They begin as linear DNA molecules packaged into bacteriophage P1 particles. These particles are injected into an E. coli strain expressing Cre recombinase. The linear P1 vector becomes circularized by recombination between two loxP sites in the vector. P1 vectors generally contain a gene for antibiotic resistance and a positive selection marker to distinguish clones containing an insert from those that do not. P1 vectors also contain a P1 plasmid replicon, which ensures only one copy of the vector is present in a cell. However, there is a second P1 replicon- called the P1 lytic replicon- that is controlled by an inducible promoter. This promoter allows the amplification of more than one copy of the vector per cell prior to DNA extraction.

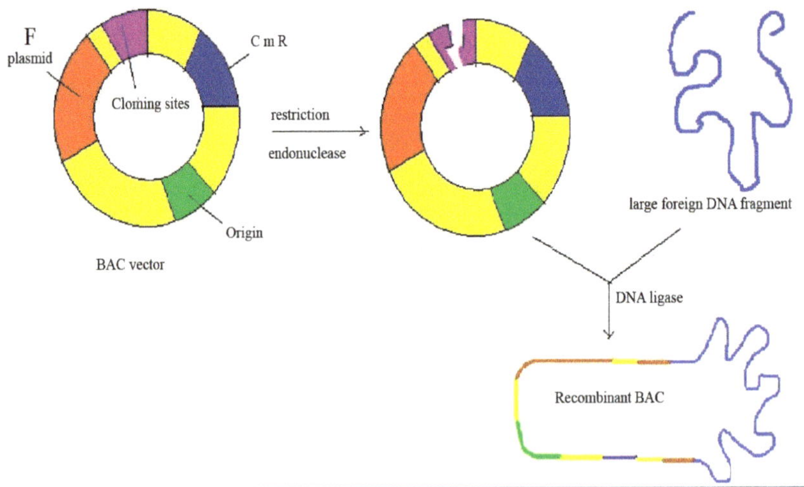

bac vector

P1 Artificial Chromosomes

P1 artificial chromosomes (PACs) have features of both P1 vectors and Bacterial Artificial Chromosomes (BACs). Similar to P1 vectors, they contain a plasmid and a lytic replicon as described above. Unlike P1 vectors, they do not need to be packaged into bacteriophage particles for transduction. Instead they are introduced into E. coli as circular DNA molecules through electroporation just as BACs are. Also similar to BACs, these are relatively harder to prepare due to a single origin of replication.

Bacterial Artificial Chromosomes

Bacterial artificial chromosomes (BACs) are circular DNA molecules, usually about 7kb in length, that are capable of holding inserts up to 300kb in size. BAC vectors contain a replicon derived from E. coli F factor, which ensures they are maintained at one copy per cell. Once an insert is ligated into a BAC, the BAC is introduced into recombination deficient strains of E. coli by electroporation. Most BAC vectors contain a gene for antibiotic resistance and also a positive selection marker. The figure to the right depicts a BAC vector being cut with a restriction enzyme, followed by the insertion of foreign DNA that is re-annealed by a ligase. Overall, this is a very stable vector, but they may be hard to prepare due to a single origin of replication just like PACs.

Yeast Artificial Chromosomes

Yeast artificial chromosomes (YACs) are linear DNA molecules containing the necessary features of an authentic yeast chromosome, including telomeres, a centromere, and an origin of replication. Large inserts of DNA can be ligated into the middle of the YAC so that there is an "arm" of the YAC on either side of the insert. The recombinant YAC is introduced into yeast by transformation; selectable markers present in the YAC allow for the identification of successful transformants. YACs can hold inserts up to 2000kb, but most YAC libraries contain inserts 250-400kb in size. Theoretically there is no upper limit on the size of insert a YAC can hold. It is the quality in the preparation of DNA used for inserts that determines the size limit. The most challenging aspect of using YAC is the fact they are prone to rearrangement.

HOW TO SELECT A VECTOR

Vector selection requires one to ensure the library made is representative of the entire genome. Any insert of the genome derived from a restriction enzyme should have an equal chance of being in the library compared to any other insert. Furthermore, recombinant molecules should contain large enough inserts ensuring the library size is able to be handled conveniently. This is particularly determined by the amount of clones needed to have in a library. The amount of clones to get a sampling of all the genes is determined by the size of the organism's genome as well as the average insert size. This is represented by the formula (also known as the Carbon and Clarke formula):

$$N = \frac{ln(1-P)}{ln(1-f)}$$

Where

N is the necessary number of recombinants

P is the desired probability that any fragment in the genome will occur at least once in the library created

f is the fractional proportion of the genome in a single recombinant

f can be further shown to be:

$$f = \frac{i}{g}$$

Where
i is the insert size
g is the genome size

Thus, increasing the insert size (by choice of vector) would allow for fewer clones needed to represent a genome. The proportion of the insert size versus the genome size represents the proportion of the respective genome in a single clone. Here is the equation with all parts considered:

$$N = \frac{ln(1-P)}{ln(1-\frac{i}{g})}$$

Vector Selection Example

The above formula can be used to determine the 99% confidence level that all sequences in a genome are represented by using a vector with an insert size of twenty thousand basepairs (such as the phage lambda vector). The genome size of the organism is three billion basepairs in this example.

$$N = \frac{ln(1-0.99)}{ln[1-\frac{2.0\times10^4 basepairs}{3.0\times10^9 basepairs}]}$$

$$N = \frac{-4.61}{-6.7\times10^{-6}}$$

$$N = 688,060 \text{ clones}$$

Thus, approximately 688,060 clones are required to ensure a 99% probability that a given DNA sequence from this three billion basepair genome will be present in a library using a vector with an insert size of twenty thousand basepairs.

Applications

After a library is created, the genome of an organism can be sequenced to elucidate how genes affect an organism or to compare similar organisms at the genome-level. The aforementioned genome-wide association studies can identify candidate genes stemming from many functional traits. Genes can be isolated through genomic libraries and used on human cell lines or animal models to further research. Furthermore, creating high-fidelity clones with accurate genome representation- and no stability issues- would contribute well as intermediates for shotgun sequencing or the study of complete genes in functional analysis.

Hierarchical Sequencing

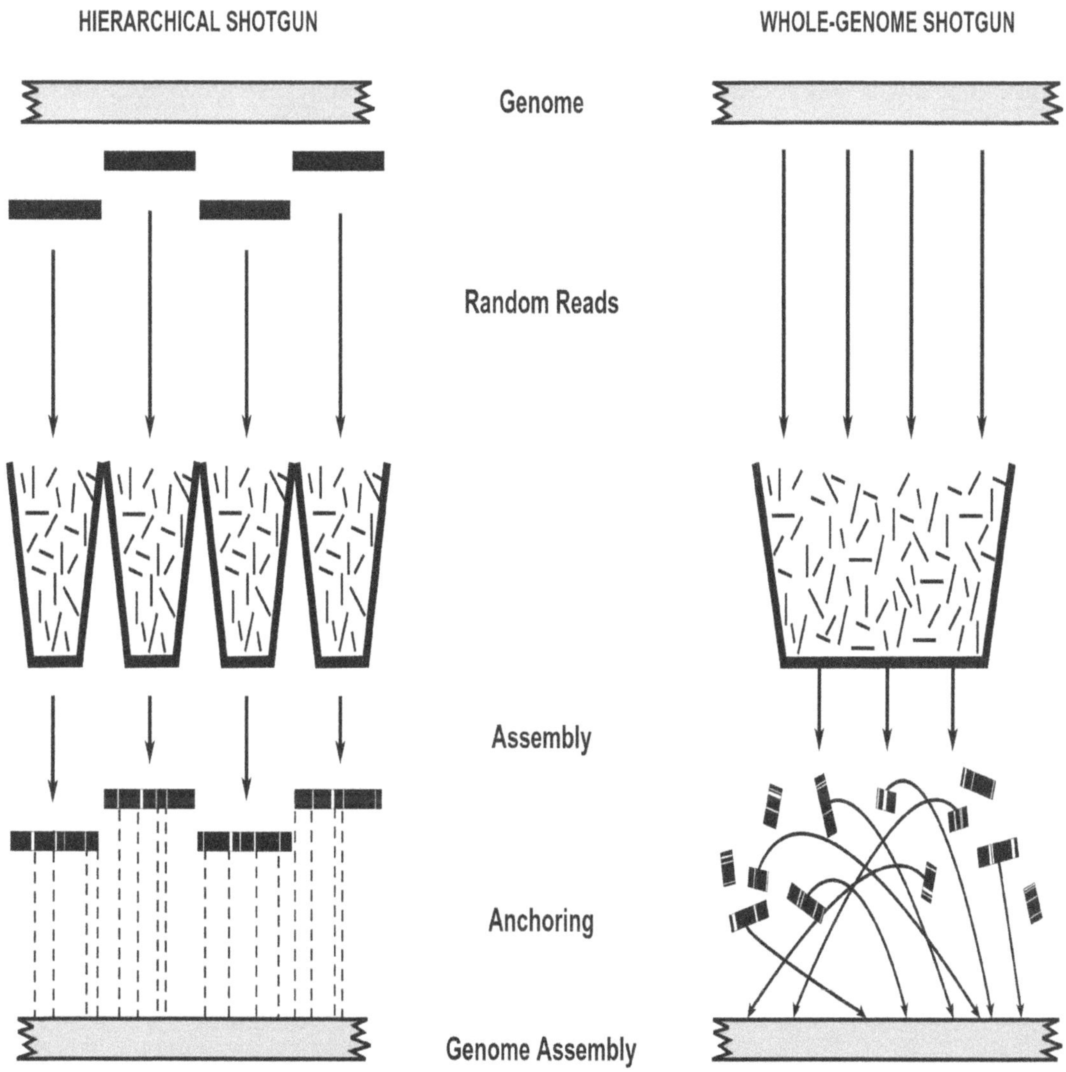

Whole genome shotgun sequencing versus Hierarchical shotgun sequencing

One major use of genomic libraries is hierarchichal shotgun sequencing, which is also called top-down, map-based or clone-by-clone sequencing. This strategy was developed in the 1980s for sequencing whole genomes before high throughput techniques for sequencing were available. Individual clones from genomic libraries can be sheared into smaller fragments, usually 500bp to 1000bp, which are more manageable for sequencing. Once a clone from a genomic library is sequenced, the sequence can be used to screen the library for other clones containing inserts which overlap with the sequenced clone. Any new overlapping clones can then be sequenced forming a contig. This technique, called chromosome walking, can be exploited to sequence entire chromosomes.

Whole genome shotgun sequencing is another method of genome sequencing that does not require a library of high-capacity vectors. Rather, it uses computer algorithms to assemble short sequence reads to cover the entire genome. Genomic libraries are often used in combination with whole genome shotgun sequencing for this reason. A high resolution map can be created by sequencing both ends of inserts from several clones in a genomic library. This map provides sequences of known distances apart, which can be used to help with the assembly of sequence reads acquired through shotgun sequencing. The human genome sequence, which was declared complete in 2003, was assembled using both a BAC library and shotgun sequencing.

Genome-Wide Association Studies

Genome-wide association studies are general applications to find specific gene targets and polymorphisms within the human race. In fact, the International HapMap project was created through a partnership of scientists and agencies from several countries to catalog and utilize this data. The goal of this project is to compare genetic sequences of different individuals to elucidate similarities and differences within chromosomal regions. Scientists from all of the participating nations are cataloging these attributes with data from populations of African, Asian, and European ancestry. Such genome-wide assessments may lead to further diagnostic and drug therapies while also helping future teams focus on orchestrating therapeutics with genetic features in mind. These concepts are already being exploited in genetic engineering. For example, a research team has actually constructed a PAC shuttle vector that creates a library representing two-fold coverage of the human genome. This could serve as an incredible resource to identify genes, or sets of genes, causing disease. Moreover, these studies can serve as a powerful way to investigate transcriptional regulation as it has been seen in the study of baculoviruses. Overall, advances in genome library construction and DNA sequencing has allowed for efficient discovery of different molecular targets.Assimilation of these features through such efficient methods can hasten the employment of novel drug candidates.

REFERENCES

1. Losick, Richard; Watson, James D.; Tania A. Baker; Bell, Stephen; Gann, Alexander; Levine, Michael W. (2008). Molecular biology of the gene. San Francisco: Pearson/Benjamin Cummings. ISBN 0-8053-9592-X.
2. Russell, David W.; Sambrook, Joseph (2001). Molecular cloning: a laboratory manual. Cold Spring Harbor, N.Y: Cold Spring Harbor Laboratory. ISBN 0-87969-577-3.
3. Hartwell, Leland (2008). Genetics: from genes to genomes. Boston: McGraw-Hill Higher Education. ISBN 0-07-284846-4.
4. Muse, Spencer V.; Gibson, Greg (2004). A primer of genome science. Sunderland, Mass: Sinauer Associates. ISBN 0-87893-232-1.
5. Henry RJ, Edwards M, Waters DL, et al. (November 2012). "Application of large-scale sequencing to marker discovery in plants". J. Biosci. 37 (5): 829–41.

doi:10.1007/s12038-012-9253-z. PMID 23107919.

6. Sanger F, Air GM, Barrell BG, et al. (February 1977). "Nucleotide sequence of bacteriophage phi X174 DNA". Nature. 265 (5596): 687–95. doi:10.1038/265687a0. PMID 870828.
7. Menon R, Farina C (2011). "Shared molecular and functional frameworks among five complex human disorders: a comparative study on interactomes linked to susceptibility genes". PLoS ONE. 6 (4): e18660. doi:10.1371/journal.pone.0018660. PMC 3080867 . PMID 21533026.
8. Cichon S, Mühleisen TW, Degenhardt FA, et al. (March 2011). "Genome-wide association study identifies genetic variation in neurocan as a susceptibility factor for bipolar disorder". Am. J. Hum. Genet. 88 (3): 372–81. doi:10.1016/j.ajhg.2011.01.017. PMC 3059436 . PMID 21353194.
9. Yoo EY, Kim S, Kim JY, Kim BD (August 2001). "Construction and characterization of a bacterial artificial chromosome library from chili pepper". Mol. Cells. 12 (1): 117–20. PMID 11561720.
10. Osoegawa K, de Jong PJ, Frengen E, Ioannou PA (May 2001). "Construction of bacterial artificial chromosome (BAC/PAC) libraries". Curr Protoc Hum Genet. Chapter 5: Unit 5.15. doi:10.1002/0471142905.hg0515s21. PMID 18428289.
11. John R. McCarrey; Williams, Steven J.; Barton E. Slatko (2006). Laboratory investigations in molecular biology. Boston: Jones and Bartlett Publishers. ISBN 0-7637-3329-6.
12. Peterson, Daniel; Jeffrey Tomkins; David Frisch (2000). "Construction of Plant Bacterial Artificial Chromosome (BAC) Libraries: An Illustrated Guide". Journal of Agricultural Genomics. 5.
13. Kim UJ, Birren BW, Slepak T, et al. (June 1996). "Construction and characterization of a human bacterial artificial chromosome library". Genomics. 34 (2): 213–8. doi:10.1006/geno.1996.0268. PMID 8661051.
14. "Cloning Genomic DNA". University College London. Retrieved 13 March 2013.
15. http://users.wmin.ac.uk/~redwayk/lectures/gene_libraries.htm
16. Blaber, Michael. "Genomic Libraries". Retrieved 1 April 2013.
17. Fuesler J, Nagahama Y, Szulewski J, Mundorff J, Bireley S, Coren JS (April 2012). "An arrayed human genomic library constructed in the PAC shuttle vector pJCPAC-Mam2 for genome-wide association studies and gene therapy". Gene. 496 (2): 103–9. doi:10.1016/j.gene.2012.01.011. PMC 3488463 . PMID 22285925.
18. Pareek CS, Smoczynski R, Tretyn A (November 2011). "Sequencing technologies and genome sequencing". J. Appl. Genet. 52 (4): 413–35. doi:10.1007/s13353-011-0057-x. PMC 3189340 . PMID 21698376.
19. Pennisi E (April 2003). "Human genome. Reaching their goal early, sequencing labs celebrate". Science. 300 (5618): 409. doi:10.1126/science.300.5618.409. PMID 12702850.
20. "HapMap Homepage".

21. Chen Y, Lin X, Yi Y, Lu Y, Zhang Z (2009). "Construction and application of a baculovirus genomic library". Z. Naturforsch. C. 64 (7-8): 574–80. doi:10.1515/znc-2009-7-817. PMID 19791511.

CHAPTER 9

Sequence Analysis

Chapter Contents

INTRODUCTION

In bioinformatics, **sequence analysis** is the process of subjecting a DNA, RNA or peptide sequence to any of a wide range of analytical methods to understand its features, function, structure, or evolution. Methodologies used include sequence alignment, searches against biological databases, and others. Since the development of methods of high-throughput production of gene and protein sequences, the rate of addition of new sequences to the databases increased exponentially. Such a collection of sequences does not, by itself, increase the scientist's understanding of the biology of organisms. However, comparing these new sequences to those with known functions is a key way of understanding the biology of an organism from which the new sequence comes. Thus, sequence analysis can be used to assign function to genes and proteins by the study of the similarities between the compared sequences. Nowadays, there are many tools and techniques that provide the sequence comparisons (sequence alignment) and analyze the alignment product to understand its biology.

Sequence analysis in molecular biology includes a very wide range of relevant topics:

- The comparison of sequences in order to find similarity, often to infer if they are related (homologous)
- Identification of intrinsic features of the sequence such as active sites, post translational modification sites, gene-structures, reading frames, distributions of introns and exons and regulatory elements
- Identification of sequence differences and variations such as point mutations and single nucleotide polymorphism (SNP) in order to get the genetic marker.
- Revealing the evolution and genetic diversity of sequences and organisms
- Identification of molecular structure from sequence alone

In chemistry, sequence analysis comprises techniques used to determine the sequence of a polymer formed of several monomers. In molecular biology and genetics, the same process is called simply «sequencing».

In marketing, sequence analysis is often used in analytical customer relationship management applications, such as NPTB models (Next Product to Buy).

In sociology, sequence methods are increasingly used to study life-course and career trajectories, patterns of organizational and national development, conversation and interaction structure, and the problem of work/family synchrony. This body of research has given rise to the emerging subfield of social sequence analysis.

Since the very first sequences of the insulin protein was characterised by Fred Sanger in 1951 biologists have been trying to use this knowledge to understand the function of molecules. He also contributed to DNA sequence. Not only he and his colleague's successes sequence the first DNA-based genome] . The method used in this study, which is called "Sanger method" or Sanger sequencing, was a milestone in sequencing long strand molecule such as DNA and this method was eventually used in human genome project. According to Michael Levitt, sequence analysis was born in the period from 1969-1977. In 1969 the analysis of sequences of transfer RNAs were used to infer residue interactions from correlated changes in the nucleotide sequences giving rise to a model of the tRNA secondary structure. In 1970, Saul B. Needleman and Christian D. Wunsch published the first computer algorithm for aligning two sequences. Over this time developments in obtaining nucleotide sequence greatly improved leading to the publication of the first complete genome of a bacteriophage in 1977. Robert Holley and his team in Cornell University was believed to be the first to sequence RNA molecule.

SEQUENCE ALIGNMENT

```
HglK       78  117  CDLRFANLEGADLTDANLIAASLHKSNLRRANLCRATLNR
HglK      118  157  CNLSEADLTESDANEALFCQAVFTEVEAHGLRLYRAKVSQ
HglK      158  197  AQLMGAHLHQAYAPEADFSAVAAIAVDLRWANLRKTNFRG
HglK      203  242  GNFRGANLTQADFTGANLKGANLRGANLVGTNLQRADLSD
spkB      454  493  QNLVGIVLAKAFVPGINCYQANLTNANFEQAELTRADFGK
spkB      504  543  ANLSDAYFGYADLRGADLRGANLNGVNFKYANLQGANFSG
slr1819    20   59  LKLPGINLEAADLIGIVLNEADLRGANLLFCYLNRANLGQ
slr1819    60   99  ANLVAANLSGASLNQADLAGADLRSANFHGAMLQGAILRD
slr1819   100  139  SDMTLATLQDTNLIGADLRGADLSGATLTGACLRGANMRQ
slr1819   151  190  AILGRADLQGANMKGVDLSRADLSYANLKEANLRDVDLRK
slr1819   191  230  ADLSYANLKGALLTDANLSGAKLNGADLQNANLMRAKISE
slr1819   231  270  AEMTAVNCQGAIMTHVNLNRTNLTGSNLSFTRMNSADLSR
slr1819   271  310  ANLTKANLQEAELIEAFFARANLTEANFINANLVRADLMS
consensus           A L  A L  A L  A L  A L  A L  A L  A L
```

Example multiple sequence alignment

There are millions of protein and nucleotide sequences known. These sequences fall into many groups of related sequences known as protein families or gene families. Relationships between these sequences are usually discovered by aligning them together and assigning this alignment a score. There are two main types of sequence alignment. Pair-wise sequence alignment only compares two sequences at a time and multiple sequence alignment compares many sequences in one go. Two important algorithms for aligning pairs of sequences are the Needleman-Wunsch algorithm and the Smith-Waterman algorithm. Popular tools for sequence alignment include:

- Pair-wise alignment - BLAST
- Multiple alignment - ClustalW, PROBCONS, MUSCLE, MAFFT, and T-Coffee.

A common use for pairwise sequence alignment is to take a sequence of interest and compare it to all known sequences in a database to identify homologous sequences. In general the matches in the database are ordered to show the most closely related sequences first followed by sequences with diminishing similarity. These matches are usually reported with a measure of statistical significance such as an Expectation value.

Profile Comparison

In 1987, Michael Gribskov, Andrew McLachlan, and David Eisenberg introduced the method of profile comparison for identifying distant similarities between proteins. Rather than using a single sequence, profile methods use a multiple sequence alignment to encode a profile which contains information about the conservation level of each residue. These profiles can then be used to search collections of sequences to find sequences that are related. Profiles are also known as Position Specific Scoring Matrices (PSSMs). In 1993, a probabilistic interpretation of profiles was introduced by David Haussler and colleagues using hidden Markov models. These models have become known as profile-HMMs.

In recent years, methods have been developed that allow the comparison of profiles directly to each other. These are known as profile-profile comparison methods.

SEQUENCE ASSEMBLY

Sequence assembly refers to the reconstruction of a DNA sequence by aligning and merging small DNA fragments. It is an integral part of modern DNA sequencing. Since presently-available DNA sequencing technologies are ill-suited for reading long sequences, large pieces of DNA (such as genomes) are often sequenced by (1) cutting the DNA into small pieces, (2) reading the small fragments, and (3) reconstituting the original DNA by merging the information on various fragment.

Recently sequencing multiple species at one time is one of the top research target. Metagenomics is studying microbial communities directly obtained from environment. Different from cultured microorganism from lab, the wild sample usually contains dozens, sometimes even thousands types of microorganisms from their original habitats. Recovering the original genomes is a real challenging work. Most recently Projects:

- Global Ocean survey (GOS)
- Data Download
- Human Microbiome Project (HMP)
- Data Download
- Earth Microbiome Project (EMP)

Gene Prediction

Gene prediction or gene finding refers to the process of identifying the regions of genomic DNA that encode genes. This includes protein-coding genes as well as RNA genes, but may also include prediction of other functional elements such as regulatory regions. Gene finding is one of the first and most important steps in understanding the genome of a species once it has been sequenced. In general the prediction of bacterial genes is significantly simpler and more accurate than the prediction of genes in eukaryotic species that usually have complex intron/exon patterns.Identifying genes in long sequences remains a problem, especially when the number of genes is unknown. Hidden markov model can be part of the solutions. Machine learning has been play a significant role in predicting the sequence of transcription factors. Traditional sequencing analyzing used focused on the statistical parameters of nucleotide sequence itself (The most common programs used are listed in Table 4.1). Another way is identifying homologous sequence based on other known gene sequence(Tools see Table 4.3). Those two methods are both focusing on sequence. However, nowadays the shape feature of these molecules such as DNA and protein have also been studied and proposed to have an equivalent influence on the behaviors of these molecular as the sequence, if not higher.

PROTEIN STRUCTURE PREDICTION

The 3D structures of molecules are of great importance to their functions in nature. Since structural prediction of large molecules at an atomic level is largely intractable problem, some biologists introduced ways to predict 3D structure at a primary sequence level. This includes biochemical or statistical analysis of amino acid residues in local regions and structural inference from homologs (or other potentially related proteins) with known 3D structures.

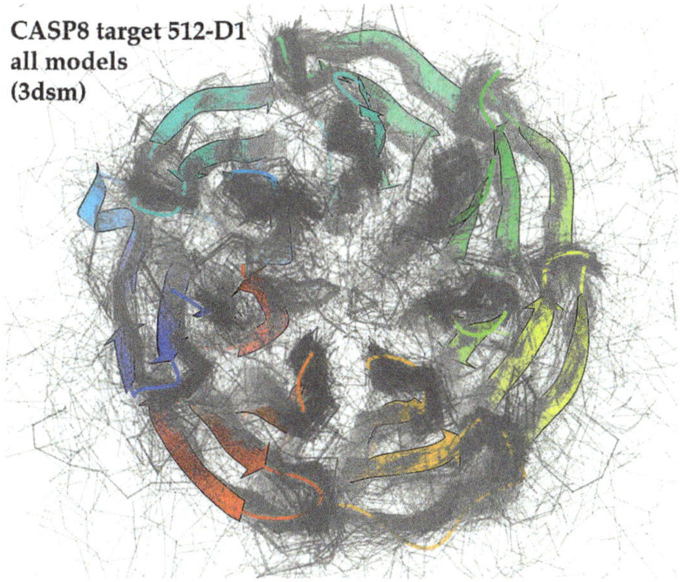

Target protein structure (3dsm, shown in ribbons), with Calpha backbones (in gray) of 354 predicted models for it submitted in the CASP8 structure-prediction experiment.

There have been a large number of diverse approaches to solve the structure prediction problem. In order to determine which methods were most effective a structure prediction competition was founded called CASP (Critical Assessment of Structure Prediction).

Methodology

The tasks that lie in the space of sequence analysis are often non-trivial to resolve and require the use of relatively complex approaches. Of the many types of methods used in practice, the most popular include:

- tools
- DNA patterns
- Dynamic programming
- Artificial Neural Network
- Hidden Markov Model
- Support Vector Machine
- Clustering
- Bayesian Network
- Regression Analysis
- Sequence mining
- Alignment-free sequence analysis

Sequence Assembly

In bioinformatics, sequence assembly refers to aligning and merging fragments from a longer DNA sequence in order to reconstruct the original sequence. This is needed as DNA sequencing technology cannot read whole genomes in one go, but rather reads small pieces of between 20 and 30000 bases, depending on the technology used. Typically the short fragments, called reads, result from shotgun sequencing genomic DNA, or gene transcript (ESTs).

The problem of sequence assembly can be compared to taking many copies of a book, passing each of them through a shredder with a different cutter, and piecing the text of the book back together just by looking at the shredded pieces. Besides the obvious difficulty of this task, there are some extra practical issues: the original may have many repeated paragraphs, and some shreds may be modified during shredding to have typos. Excerpts from another book may also be added in, and some shreds may be completely unrecognizable.

Genome Assemblers

The first sequence assemblers began to appear in the late 1980s and early 1990s as variants of simpler sequence alignment programs to piece together vast quantities of fragments generated by automated sequencing instruments called DNA sequencers. As the sequenced organisms grew in size and complexity (from small viruses over plasmids to bacteria and finally eukaryotes), the assembly programs used in these genome projects needed increasingly sophisticated strategies to handle:

- terabytes of sequencing data which need processing on computing clusters;
- identical and nearly identical sequences (known as *repeats*) which can, in the worst case, increase the time and space complexity of algorithms exponentially;
- errors in the fragments from the sequencing instruments, which can confound assembly.

Faced with the challenge of assembling the first larger eukaryotic genomes—the fruit fly Drosophila melanogaster in 2000 and the human genome just a year later,—scientists developed assemblers like Celera Assembler and Arachne able to handle genomes of 100-300 million base pairs. Subsequent to these efforts, several other groups, mostly at the major genome sequencing centers, built large-scale assemblers, and an open source effort known as AMOS was launched to bring together all the innovations in genome assembly technology under the open source framework.

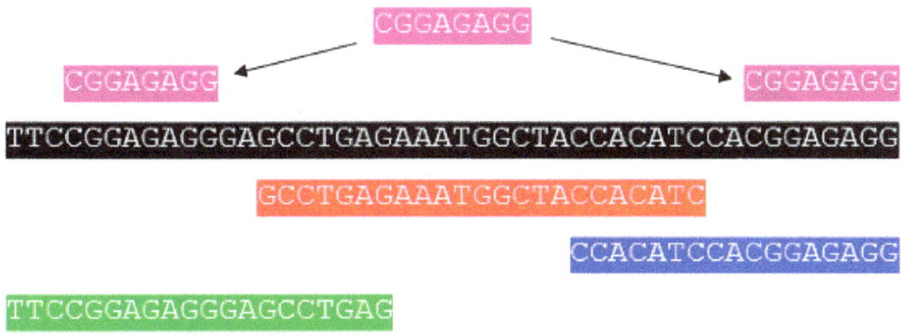

Sample sequence showing how a sequence assembler would take fragments and match by overlaps. Image also shows the potential problem of repeats in the sequence.

EST Assemblers

Expressed Sequence Tag or EST assembly differs from genome assembly in several ways. The sequences for EST assembly are the transcribed mRNA of a cell and represent only a subset of the whole genome. At a first glance, underlying algorithmical problems differ between genome and EST assembly. For instance, genomes often have large amounts of repetitive sequences, mainly in the inter-genic parts. Since ESTs represent gene transcripts, they will not contain these repeats. On the other hand, cells tend to have a certain number of genes that are constantly expressed in very high numbers (housekeeping genes), which again leads to the problem of similar sequences present in high numbers in the data set to be assembled.

Furthermore, genes sometimes overlap in the genome (sense-antisense transcription), and should ideally still be assembled separately. EST assembly is also complicated by features like (cis-) alternative splicing, trans-splicing, single-nucleotide polymorphism, and post-transcriptional modification.

De-Novo vs. Mapping Assembly

In sequence assembly, two different types can be distinguished:

- de-novo: assembling short reads to create full-length (sometimes novel) sequences (see de novo transcriptome assembly)
- mapping: assembling reads against an existing backbone sequence, building a sequence that is similar but not necessarily identical to the backbone sequence

In terms of complexity and time requirements, de-novo assemblies are orders of magnitude slower and more memory intensive than mapping assemblies. This is mostly due to the fact that the assembly algorithm needs to compare every read with every other read (an operation that has a naive time complexity of $O(n^2)$; using a hash this can be reduced significantly). Referring to the comparison drawn to shredded books in the introduction: while for mapping assemblies one would have a very similar book as template (perhaps with the names of the main characters and a few locations changed), the de-novo assemblies are more hardcore in a sense as one would not know beforehand whether this would become a science book, a novel, a catalogue, or even several books. Also, every shred would be compared with every other shred.

Influence of Technological Changes

The complexity of sequence assembly is driven by two major factors: the number of fragments and their lengths. While more and longer fragments allow better identification of sequence overlaps, they also pose problems as the underlying algorithms show quadratic or even exponential complexity behaviour to both number of fragments and their length. And while shorter sequences are faster to align, they also complicate the layout phase of an assembly as shorter reads are more difficult to use with repeats or near identical repeats.

In the earliest days of DNA sequencing, scientists could only gain a few sequences of short length (some dozen bases) after weeks of work in laboratories. Hence, these sequences could be aligned in a few minutes by hand.

In 1975, the *Dideoxy termination* method (AKA *Sanger sequencing*) was invented and until shortly after 2000, the technology was improved up to a point where fully automated machines could churn out sequences in a highly parallelised mode 24 hours a day. Large genome centers around the world housed complete farms of these sequencing machines, which in turn led to the necessity of assemblers to be optimised for sequences from whole-genome shotgun sequencing projects where the reads

- are about 800–900 bases long
- contain sequencing artifacts like sequencing and cloning vectors
- have error rates between 0.5 and 10%

With the Sanger technology, bacterial projects with 20,000 to 200,000 reads could easily be assembled on one computer. Larger projects, like the human genome with approximately 35 million reads, needed large computing farms and distributed computing.

By 2004 / 2005, pyrosequencing had been brought to commercial viability by 454 Life Sciences. This new sequencing method generated reads much shorter than those of Sanger sequencing: initially about 100 bases, now 400-500 bases. Its much higher throughput and lower cost (compared to Sanger sequencing) pushed the adoption of this technology by genome centers, which in turn pushed development of sequence assemblers that could efficiently handle the read sets. The sheer amount of data coupled with technology-specific error patterns in the reads delayed development of assemblers; at the beginning in 2004 only the Newbler assembler from 454 was available. Released in mid-2007, the hybrid version of the MIRA assembler by Chevreux et al. was the first freely available assembler that could assemble 454 reads as well as mixtures of 454 reads and Sanger reads. Assembling sequences from different sequencing technologies was subsequently coined *hybrid assembly*.

From 2006, the Illumina (previously Solexa) technology has been available and can generate about 100 million reads per run on a single sequencing machine. Compare this to the 35 million reads of the human genome project which needed several years to be produced on hundreds of sequencing machines. Illumina was initially limited to a length of only 36 bases, making it less suitable for de novo assembly (such as de novo transcriptome assembly), but newer iterations of the technology achieve read lengths above 100 bases from both ends of a 3-400bp clone. Announced at the end of 2007, the SHARCGS assembler by Dohm et al. was the first published assembler that was used for an assembly with Solexa reads. It was quickly followed by a number of others.

Later, new technologies like SOLiD from Applied Biosystems, Ion Torrent and SMRT were released and new technologies (e.g. Nanopore sequencing) continue to emerge.

Greedy Algorithm

Given a set of sequence fragments the object is to find the shortest common supersequence.

- Calculate pairwise alignments of all fragments.
- Choose two fragments with the largest overlap.
- Merge chosen fragments.
- Repeat step 2 and 3 until only one fragment is left.

The result need not be an optimal solution to the problem.

Available Assemblers

The following table lists assemblers that have a de-novo assembly capability on at least one of the supported technologies.

Name	Type	Technologies	Author	Presented / Last updated	Licence*	Homepage
ABySS	(large) genomes	Solexa, SOLiD	Simpson, J. et al.	2008 / 2014	NC-A	link
ALLPATHS-LG	(large) genomes	Solexa, SOLiD	Gnerre, S. et al.	2011	OS	link
AMOS	genomes	Sanger, 454	Salzberg, S. et al.	2002? / 2011	OS	link
Arapan-M	Medium Genomes (e.g. E.coli)	All	Sahli, M. & Shibuya, T.	2011 / 2012	OS	link
Arapan-S	Small Genomes (Viruses and Bacteria)	All	Sahli, M. & Shibuya, T.	2011 / 2012	OS	link
Celera WGA Assembler / CABOG	(large) genomes	Sanger, 454, Solexa	Myers, G. et al.; Miller G. et al.	2004 / 2015	OS	link
CLC Genomics Workbench & CLC Assembly Cell	genomes	Sanger, 454, Solexa, SOLiD	CLC bio	2008 / 2010 / 2014	C	link
Cortex	genomes	Solexa, SOLiD	Iqbal, Z. *et al.*	2011	OS	link

Name	Type	Technologies	Author	Presented / Last updated	Licence*	Homepage
DBG2OLC	(large) genomes	Illumina, PacBio, Oxford Nanopore	Ye, C. et al	2014/2016	OS	link
DNA Baser Assembler	(small) genomes	Sanger, 454	Heracle BioSoft SRL	04.2016	C	www. DnaBaser.com
DNA Dragon	genomes	Illumina, SOLiD, Complete Genomics, 454, Sanger	SequentiX	2011	C	link
DNAnexus	genomes	Illumina, SOLiD, Complete Genomics	DNAnexus	2011	C	link
DNASTAR Lasergene Genomics Suite	(large) genomes, exomes, transcriptomes, metagenomes, ESTs	Illumina, ABI SOLiD, Roche 454, Ion Torrent, Solexa, Sanger	DNASTAR	2007 / 2016	C	link
Edena	genomes	Illumina	D. Hernandez, P. François, L. Farinelli, M. Osteras, and J. Schrenzel.	2008/2013	OS	link
Euler	genomes	Sanger, 454 (,Solexa ?)	Pevzner, P. et al.	2001 / 2006?	(C / NC-A?)	link
Euler-sr	genomes	454, Solexa	Chaisson, MJ. et al.	2008	NC-A	link
Fermi	(large) genomes	Illumina	Li, H.	2012	OS	link
Forge	(large) genomes, EST, metagenomes	454, Solexa, SOLID, Sanger	Platt, DM, Evers, D.	2010	OS	link
Geneious	genomes	Sanger, 454, Solexa, Ion Torrent, Complete Genomics, PacBio, Oxford Nanopore, Illumina	Biomatters Ltd	2009 / 2013	C	link
Graph Constructor	(large) genomes	Sanger, 454, Solexa, SOLiD	Convey Computer Corporation	2011	C	link
HINGE	genomes	PacBio/Oxford Nanopore	Kamath, Shomorony, Xia et. al.	2016	OS	Software, Paper, Analyses
IDBA (Iterative De Bruijn graph short read Assembler)	(large) genomes	Sanger,454,Solexa	Yu Peng, Henry C. M. Leung, Siu-Ming Yiu, Francis Y. L. Chin	2010	(C / NC-A?)	link
LIGR Assembler (derived from TIGR Assembler)	genomic	Sanger	-	2009/ 2012	OS	link

Name	Type	Technologies	Author	Presented / Last updated	Licence*	Homepage
MaSuRCA (Maryland Super Read - Celera Assembler)	(large) genomes	Sanger, Illumina, 454	Aleksey Zimin, Guillaume Marçais, Daniela Puiu, Michael Roberts, Steven L. Salzberg, James A. Yorke	2012 / 2013	OS	link
MIRA (Mimicking Intelligent Read Assembly)	genomes, ESTs	Sanger, 454, Solexa	Chevreux, B.	1998 / 2014	OS	link
NextGENe	(small genomes?)	454, Solexa, SOLiD	Softgenetics	2008	C	link
Newbler	genomes, ESTs	454, Sanger	454/Roche	2004/2012	C	link
PADENA	genomes	454, Sanger	454/Roche	2010	OS	link
PASHA	(large) genomes	Illumina	Liu, Schmidt, Maskell	2011	OS	link
Phrap	genomes	Sanger, 454, Solexa	Green, P.	1994 / 2008	C / NC-A	link
TIGR Assembler	genomic	Sanger	-	1995 / 2003	OS	link
Trinity	Transcriptomes	short reads (paired, oriented, mixed) Illumina, 454, Solid,...	Grabher, MG et al.	2011/2016	OS	https://github.com/trinityrnaseq/trinityrnaseq/wiki
Ray	genomes	Illumina, mix of Illumina and 454, paired or not	Sébastien Boisvert, François Laviolette & Jacques Corbeil.	2010	OS GNU General Public License	link
Sequencher	genomes	traditional and next generation sequence data	Gene Codes Corporation	1991 / 2009 / 2011	C	link
SGA	(large) genomes	Illumina, Sanger (Roche 454?, Ion Torrent?)	Simpson, J.T. et al.	2011 / 2012	OS	link
SHARCGS	(small) genomes	Solexa	Dohm et al.	2007 / 2007	OS	link
SOPRA	genomes	Illumina, SOLiD, Sanger, 454	Dayarian, A. et al.	2010 / 2011	OS	link
SparseAssembler	(large) genomes	Illumina, 454, Ion torrent	Ye, C. et al.	2012 / 2012	OS	link
SSAKE	(small) genomes	Solexa (SOLiD? Helicos?)	Warren, R. et al.	2007 / 2014	OS	link
SOAPdenovo	genomes	Solexa	Luo, R. et al.	2009 / 2013	OS	link
SPAdes	(small) genomes, single-cell	Illumina, Solexa, Sanger, 454, Ion Torrent, PacBio, Oxford Nanopore	Bankevich, A et al.	2012 / 2015	OS	link
Staden gap4 package	BACs (, small genomes?)	Sanger	Staden et al.	1991 / 2008	OS	link
Taipan	(small) genomes	Illumina	Schmidt, B. *et al.*	2009 / 2009	OS	link

Name	Type	Technologies	Author	Presented / Last updated	Licence*	Homepage
VCAKE	(small) genomes	Solexa (SOLiD?, Helicos?)	Jeck, W. et al.	2007 / 2009	OS	link
Phusion assembler	(large) genomes	Sanger	Mullikin JC, *et al.*	2003 / 2006	OS	link
Quality Value Guided SRA (QSRA)	genomes	Sanger, Solexa	Bryant DW, *et al.*	2009 / 2009	OS	link
Velvet	(small) genomes	Sanger, 454, Solexa, SOLiD	Zerbino, D. et al.	2007 / 2011	OS	link

REFERENCES

1. Myers, E. W.; Sutton, GG; Delcher, AL; Dew, IM; Fasulo, DP; Flanigan, MJ; Kravitz, SA; Mobarry, CM; et al. (March 2000). "A whole-genome assembly of Drosophila". Science. 287 (5461): 2196–204. doi:10.1126/science.287.5461.2196. PMID 10731133.
2. Batzoglou, S.; Jaffe, DB; Stanley, K; Butler, J; Gnerre, S; Mauceli, E; Berger, B; Mesirov, JP; Lander, ES (January 2002). "ARACHNE: a whole-genome shotgun assembler". Genome Research. 12 (1): 177–89. doi:10.1101/gr.208902. PMC 155255 . PMID 11779843.
3. AMOS page with links to various papers
4. Copy in Google groups of the post announcing MIRA 2.9.8 hybrid version in the bionet.software Usenet group
5. Dohm, J. C.; Lottaz, C.; Borodina, T.; Himmelbauer, H. (November 2007). "SHARCGS, a fast and highly accurate short-read assembly algorithm for de novo genomic sequencing". Genome Research. 17 (11): 1697–706. doi:10.1101/gr.6435207. PMC 2045152 . PMID 17908823.
6. list of software including mapping assemblers in the SeqAnswers discussion forum.
7. Kamath, Govinda M.; Shomorony, Ilan; Xia, Fei; Courtade, Thomas; Tse, David N. (1 August 2016). "HINGE: Long-Read Assembly Achieves Optimal Repeat Resolution" (PDF). biorXiv preprint.
8. Grabherr, Manfred G.; Haas, Brian J.; Yassour, Moran; Levin, Joshua Z.; Thompson, Dawn A.; Amit, Ido; Adiconis, Xian; Fan, Lin; Raychowdhury, Raktima (2011-07-01). "Full-length transcriptome assembly from RNA-Seq data without a reference genome". Nature Biotechnology. 29 (7): 644–652. doi:10.1038/nbt.1883. ISSN 1087-0156. PMC 3571712 . PMID 21572440.
9. Boisvert, Sébastien; Laviolette, François; Corbeil, Jacques (October 2010). "Ray: simultaneous assembly of reads from a mix of high-throughput sequencing technologies". Journal of Computational Biology. 17 (11): 1519–33. doi:10.1089/cmb.2009.0238. PMC 3119603 . PMID 20958248.
10. Durbin, Richard M.; Eddy, Sean R.; Krogh, Anders; Mitchison, Graeme (1998),

Biological Sequence Analysis: Probabilistic Models of Proteins and Nucleic Acids(1st ed.), Cambridge: Cambridge University Press, doi:10.2277/0521629713, ISBN 0-521-62971-3

11. Sanger F; Tuppy H (September 1951). "The amino-acid sequence in the phenylalanyl chain of insulin. I. The identification of lower peptides from partial hydrolysates". Biochem. J. 49 (4): 463–81. doi:10.1042/bj0490463. PMC 1197535 . PMID 14886310.
12. SANGER F; TUPPY H (September 1951). "The amino-acid sequence in the phenylalanyl chain of insulin. 2. The investigation of peptides from enzymic hydrolysates". Biochem. J. 49 (4): 481–90. doi:10.1042/bj0490481. PMC 1197536 . PMID 14886311.
13. Sanger, F; Nicklen, S; Coulson, AR (1977 December). "DNA sequencing with chain-terminating inhibitors". Proc Natl Acad Sci U S A. 74 (12): 441–448. doi:10.1073/pnas.74.12.5463. PMC 431765 . PMID 271968. Check date values in: |date= (help)
14. Sanger, F; Air, GM; Barrell, BG; Brown, NL; Coulson, AR; Fiddes, CA; Hutchison, CA; Slocombe, PM; Smith, M (1977 Feb 24). "Nucleotide sequence of bacteriophage phi X174 DNA". Nature. 265 (5596): 687–695. doi:10.1038/265687a0. PMID 870828. Retrieved 15 November 2016. Check date values in: |date= (help)
15. Levitt M (May 2001). "The birth of computational structural biology". Nature Structural & Molecular Biology. 8 (5): 392–3. doi:10.1038/87545. PMID 11323711.
16. Levitt M (November 1969). "Detailed molecular model for transfer ribonucleic acid". Nature. 224 (5221): 759–63. Bibcode:1969Natur.224..759L. doi:10.1038/224759a0. PMID 5361649.
17. Needleman SB; Wunsch CD (March 1970). "A general method applicable to the search for similarities in the amino acid sequence of two proteins". J. Mol. Biol. 48 (3): 443–53. doi:10.1016/0022-2836(70)90057-4. PMID 5420325.
18. Sanger F, Air GM, Barrell BG, et al. (February 1977). "Nucleotide sequence of bacteriophage phi X174 DNA". Nature. 265 (5596): 687–95. Bibcode:1977Natur.265..687S. doi:10.1038/265687a0. PMID 870828.
19. Holley, RW; Apgar, J; Everett, GA; Madison, JT; Marquisee, M; Merrill, SH; Penswick, JR; Zamir, A (1965 May 19). "Structure of a Ribonucleic Acid". Science. 147 (3664): 1462–1465. doi:10.1126/science.147.3664.1462. PMID 14263761.Check date values in: |date= (help);
20. Gribskov M; McLachlan AD; Eisenberg D (July 1987). "Profile analysis: detection of distantly related proteins". Proc. Natl. Acad. Sci. U.S.A. 84 (13): 4355–8. Bibcode:1987PNAS...84.4355G. doi:10.1073/pnas.84.13.4355. PMC 305087 . PMID 3474607.
21. Brown M; Hughey R; Krogh A; Mian IS; Sjölander K; Haussler D (1993). "Using Dirichlet mixture priors to derive hidden Markov models for protein families". Proc Int Conf Intell Syst Mol Biol. 1: 47–55. PMID 7584370.

22. Krogh A; Brown M; Mian IS; Sjölander K; Haussler D (February 1994). "Hidden Markov models in computational biology. Applications to protein modeling". J. Mol. Biol. 235 (5): 1501–31. doi:10.1006/jmbi.1994.1104. PMID 8107089.
23. Ye X; Wang G; Altschul SF (December 2011). "An assessment of substitution scores for protein profile-profile comparison". Bioinformatics. 27 (24): 3356–63. doi:10.1093/bioinformatics/btr565. PMC 3232366 . PMID 21998158.
24. Wooley, JC; Godzik, A; Friedberg, I (2010 Feb 26). "A primer on metagenomics". PLoS Comput Biol. 6 (2): e1000667. doi:10.1371/journal.pcbi.1000667. PMC 2829047 . PMID 20195499. Check date values in: |date= (help)
25. Stanke, M; Waack, S (2003 Oct 19). "Gene prediction with a hidden Markov model and a new intron submodel". Bioinformatics. 2: 215–25. PMID 14534192. Check date values in: |date= (help);
26. Alipanahi, B; Delong, A; Weirauch, MT; Frey, BJ (2015 Aug). "Predicting the sequence specificities of DNA- and RNA-binding proteins by deep learning". Nat Biotechnol. 33 (8): 831–8. doi:10.1038/nbt.3300. PMID 26213851. Check date values in: |date= (help);
27. Wooley, JC; Godzik, A; Friedberg, I (2010 Feb 26). "A primer on metagenomics". PLoS Comput Biol. 6 (2): e1000667. doi:10.1371/journal.pcbi.1000667. PMC 2829047 . PMID 20195499. Retrieved 15 November 2016. Check date values in: |date= (help)
28. Abe, N; Dror, I; Yang, L; Slattery, M; Zhou, T; Bussemaker, HJ; Rohs R, R; Mann, RS (2015 Apr 9). "Deconvolving the recognition of DNA shape from sequence". Cell. 161 (2): 307–18. doi:10.1016/j.cell.2015.02.008. PMC 4422406 . PMID 25843630. Check date values in: |date= (help);
29. Moult J; Hubbard T; Bryant SH; Fidelis K; Pedersen JT (1997). "Critical assessment of methods of protein structure prediction (CASP): round II". Proteins. Suppl 1: 2–6. doi:10.1002/(SICI)1097-0134(1997)1+<2::AID-PROT2>3.0.CO;2-T. PMID 9485489.

Index